For
Grandma and Grandpa
Love
Chris.

W9-CGU-270

MotionGraphics FILM+TV

EDITED BY
KATHLEEN ZIEGLER
NICK GRECO
TAMYE RIGGS

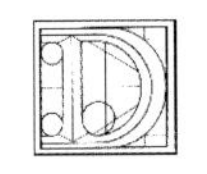

Dimensional Illustrators, Inc.

an imprint of HarperCollins International

First Published 2002 by:
HBI, an imprint of
HarperCollins Publishers
10 East 53rd Street
New York, NY 10022-5299

Distributed in the US and Canada by:
Watson-Guptill Publications
770 Broadway
New York, NY 10022-5299
Telephone: 800.451.1741 or 732.363.4511 in NJ, AK, HI
Fax: 732.363.0338

ISBN 0-8230-3141-1

Distributed throughout the rest of the world by:
HarperCollins International
10 East 53rd Street
New York, NY 10022-5299
Fax: 212.207.7654

ISBN 0-06-008759-5

Address Direct Mail Sales to:
Dimensional Illustrators, Inc.
362 Second Street Pike / # 112
Southampton, Pennsylvania 18966 USA
Telephone: 215.953.1415
Fax: 215.953.1697
Email: dimension@3dimillus.com

Printed in Hong Kong.

CREDITS

Creative Director / Creative Editor
Kathleen Ziegler Dimensional Illustrators, Inc.

Executive Editor
Nick Greco Dimensional Illustrators, Inc.

Contributing Editor
Tamye Riggs

Design and Typography
Deborah Davis Deborah Davis Design

Artwork
Page 7, 158 **©2002 Honest**

Page 8-9 **©2002 Verb.**

TABLE OF CONTENTS

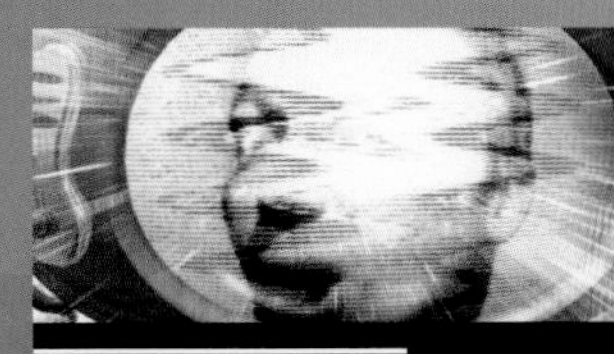

Scientific Method
Galileo Galilei

TRL Superstar:

THE POWER

FLIX EVENT

FLIX PICK

FLIX 50's

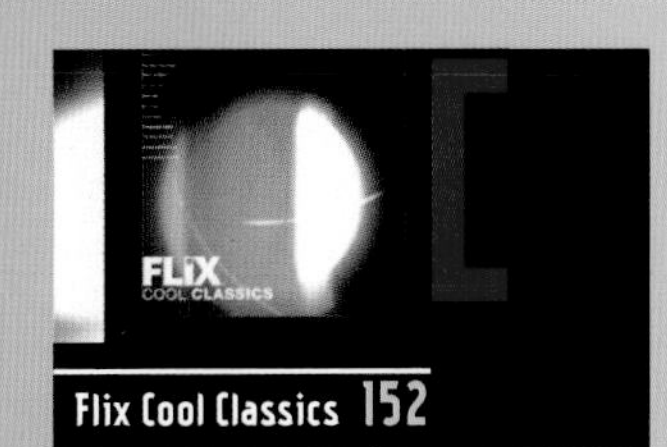
FLIX
COOL CLASSICS

VOLVO
for life
HYSTE
RICAL
BLIND
NESS

INTRODUCTION

MotionGraphics Film + TV contains an electrifying series of motion graphics developed for broadcast and film. Thirty-five of the world's most innovative studios and independent designers demonstrate the impact of fusing creativity and technology in contemporary media. Seventy-five featured projects present a dynamic overview of exciting motion work from Argentina, Australia, Canada, France, Japan, the Slovak Republic, Sweden, the United Kingdom and the United States. Live action footage, liquid effects, richly layered montages, kinetic visuals and

stunning typography are combined with lush, rhythmic soundtracks to captivate sophisticated audiences. These Film Title Sequences, TV Show Openers, Commercials, Network Identity and Promotions, Public Service Announcements, Video Installations, Short Films and Music Videos are just a sampling of the imaginative compositions displayed on the air and on the giant screen. This fantastic voyage cuts through the static, revealing a sensory feast for those who like to watch.

—Dimensional Illustrators, Inc.

The Making of Jose Cuervo Tequila

OVT was commissioned to create this sequence representing the dusty farms of Mexico where Jose Cuervo Tequila was born. The grungy logo treatment and earthy tones evoke the emotional appeal of the product. The original footage was shot on BetaSP in Mexico. Logo and type treatments were created in Adobe Illustrator, then composited in Adobe After Effects. Logos were motion-tracked to live action elements. The resulting piece entices the viewer into the world of Jose Cuervo, urging the uninitiated to partake, and the sophisticated tequila connoisseur to revisit the Cuervo experience.

Designers **Brian Check, Doug Seay** Studio **OVT Visuals, Inc.**

Point-of-Purchase Video Display **Old World Stoli** | Producer **Brian Dressel** Creative Director **Brian Dressel**

Old World Stoli

The design of this sequence from OVT Visuals is a nod to Russian Constructivism and the monolithic propaganda that emerged after Stalin's reign. Playing like old movie film found in a Russian warehouse, the piece is spiked with the energy and electricity of the post-Berlin Wall era. The logos and type treatment were inspired by old Russian posters and propaganda. The strong lines, bold text and red coloring throughout suggest the pride-rooted nobility of the Stoli brand.

DesignersBrian Dressel, Brian Check StudioOVT Visuals, Inc.

Tanqueray® № TEN

Tanqueray No. 10

The new and luxuriously designed bottle of Tanqueray No.10 is represented by the elegant 3D graphics and liquid effects throughout the sequence. The logos and text are presented as liquid textures, heightening the sophistication and grace of the world-famous gin. The jewel-toned green dominating the piece is a unifying brand identification long associated with the product. OVT's fluid and rich imagery evokes a thirst for the smooth quality of a Tanqueray martini.

Video Installation **Tanqueray No. 10** Producer **Brian Dressel** Creative Director **Brian Dressel** Designers **Brian Dressel,**

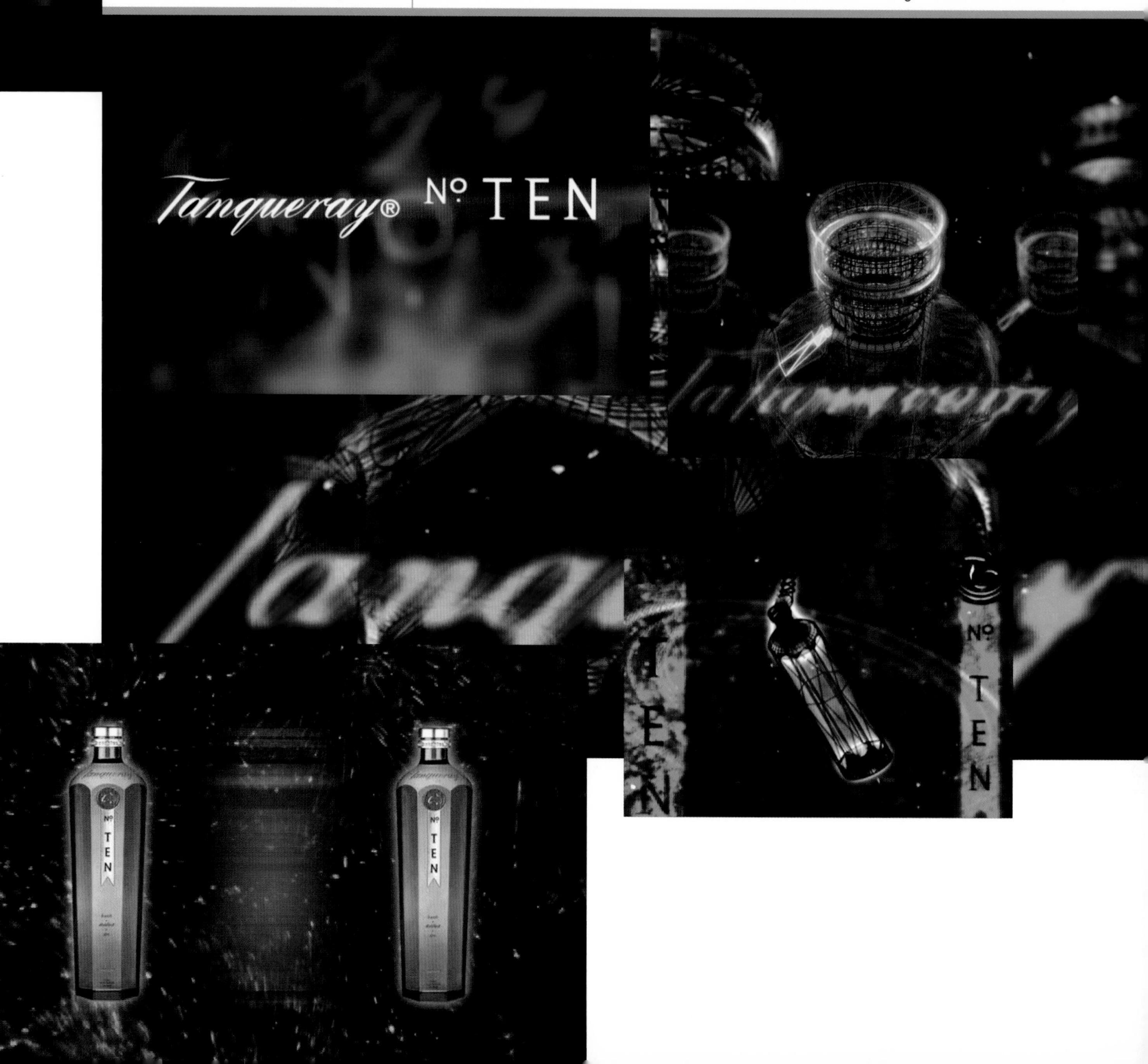

Brian Check StudioOVT Visuals, Inc.

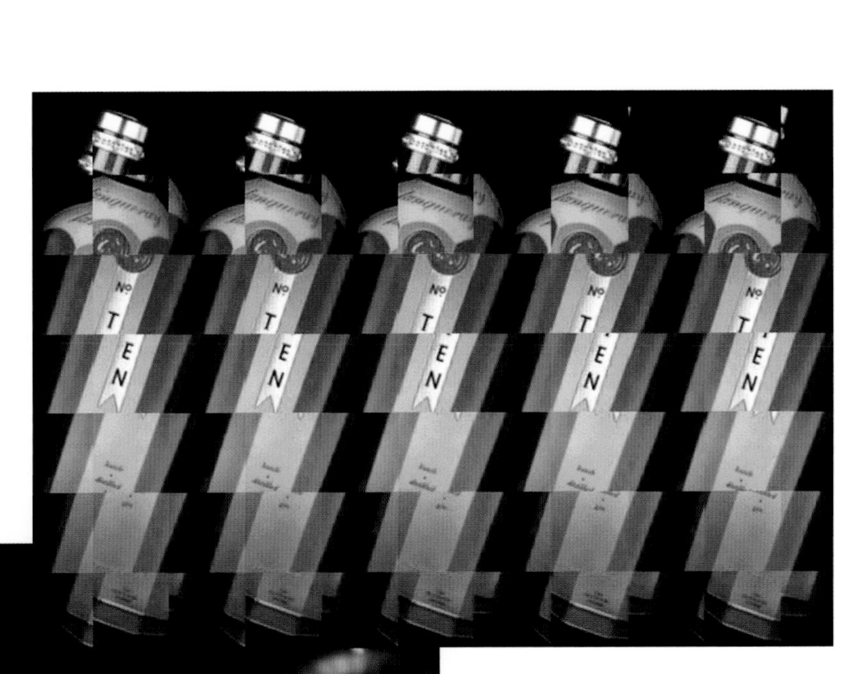

Video Installation **Night-Vision** | Producer **Brian Dressel** Creative Director **Brian Dressel** Designers **Peter Szewczyk,**

Night-Vision

This piece from Chicago-based OVT Visuals captures the electricity and kinetic energy of the modern night-club environment. The look of the Night-Vision project was inspired by artificial robot vision. Live action footage of people dancing and drinking was manipulated and composited with computer readouts, 3D wireframes and geometric shapes. The warm, dark, electric cave of a crowded nightclub serves as a metaphor for an increasingly technological society—more connected, yet more apart than ever.

Jason Voke, Brian Dressel StudioOVT Visuals, Inc.

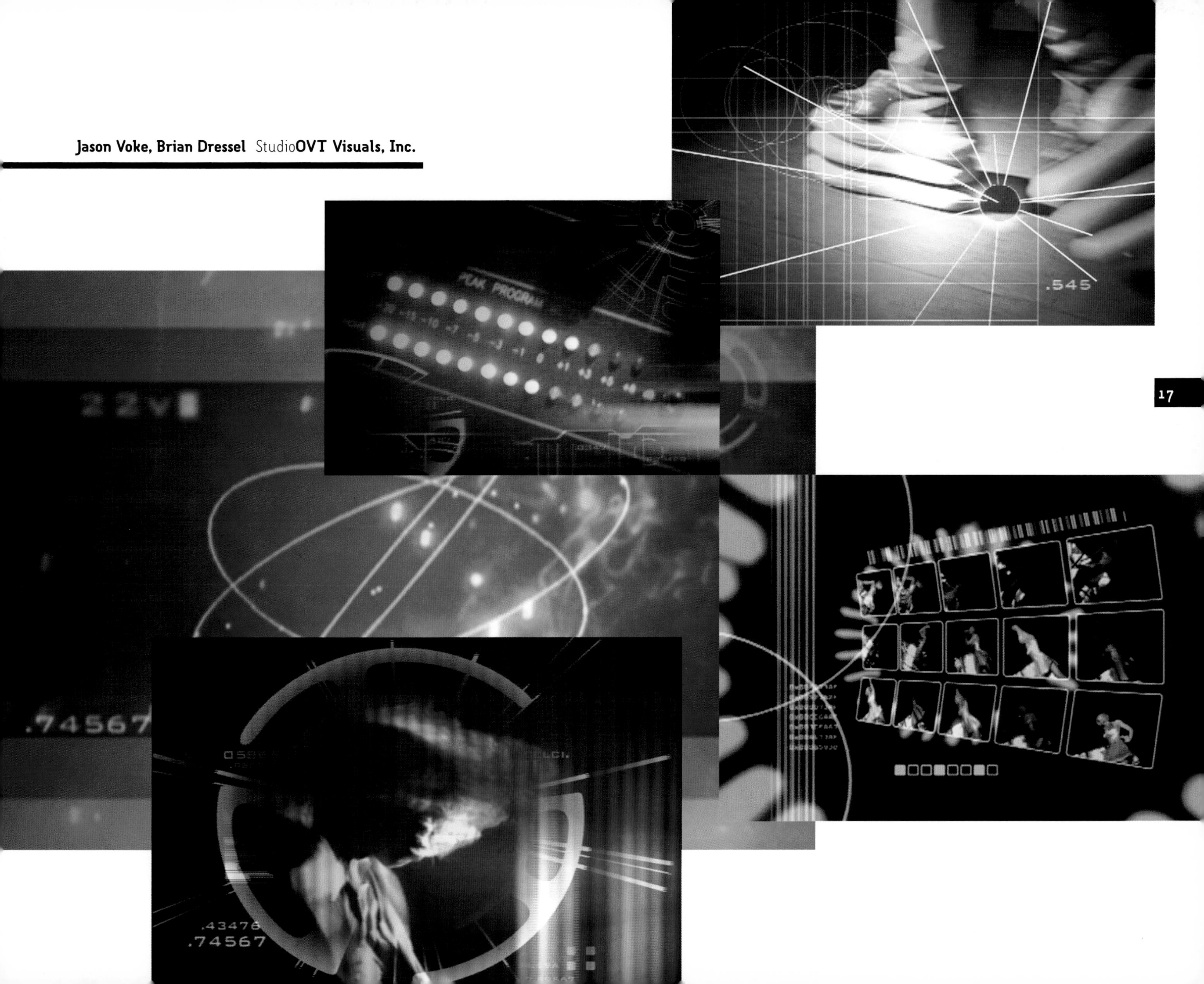

TNT Basketball Graphics

TNT's creative director spotted a killer piece on twenty2product's reel and asked them to simulate it using NBA stars. 22p recycled a project originally created for Niketown NYC, featuring a series of shuffling images of athletes revolving, slot machine-style, to a single featured athlete. This project uses a combination of two filter effects in Adobe PremiereRoll and Ghost. The studio used this secret recipe in numerous Nike projects during 1993–1995 before they worked with Adobe After Effects. Green created several color and tempo variations, then Tolson combined them with the type animations. The result is a fast-paced, exciting spotlight for the NBA's superstar players.

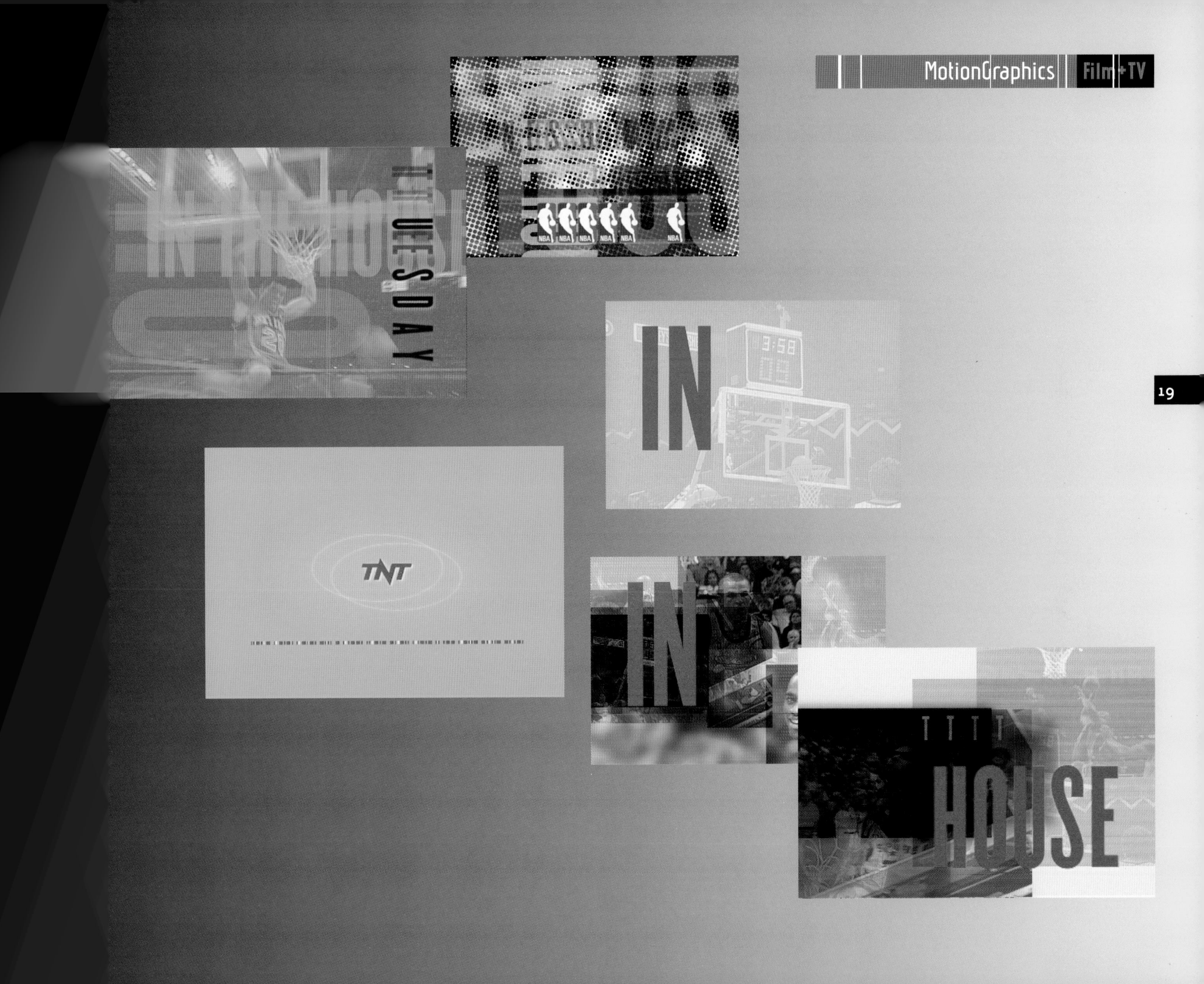

TUESDAY
IN THE HOUSE
TNT
3:58
IN
IN
HOUSE

RockShox 2000 Tradeshow Montage

This project was awarded to twenty2product while they were working on the client's website. RockShox asked 22p to put together a video to introduce the site's navigation conventions to tradeshow attendees. Tolson cut Hamish Beattie's live action to the music track, then rebuilt all the technical animations initially set up in Macromedia Flash, in Adobe After Effects. Simultaneously, Green sequenced the website's initial Photoshop layouts into short "burst-assembly" animations in Adobe After Effects. This energetic and educational promo piece added to the buzz surrounding the launch of the new RockShox site.

RockShox 2000 Tradeshow Montage

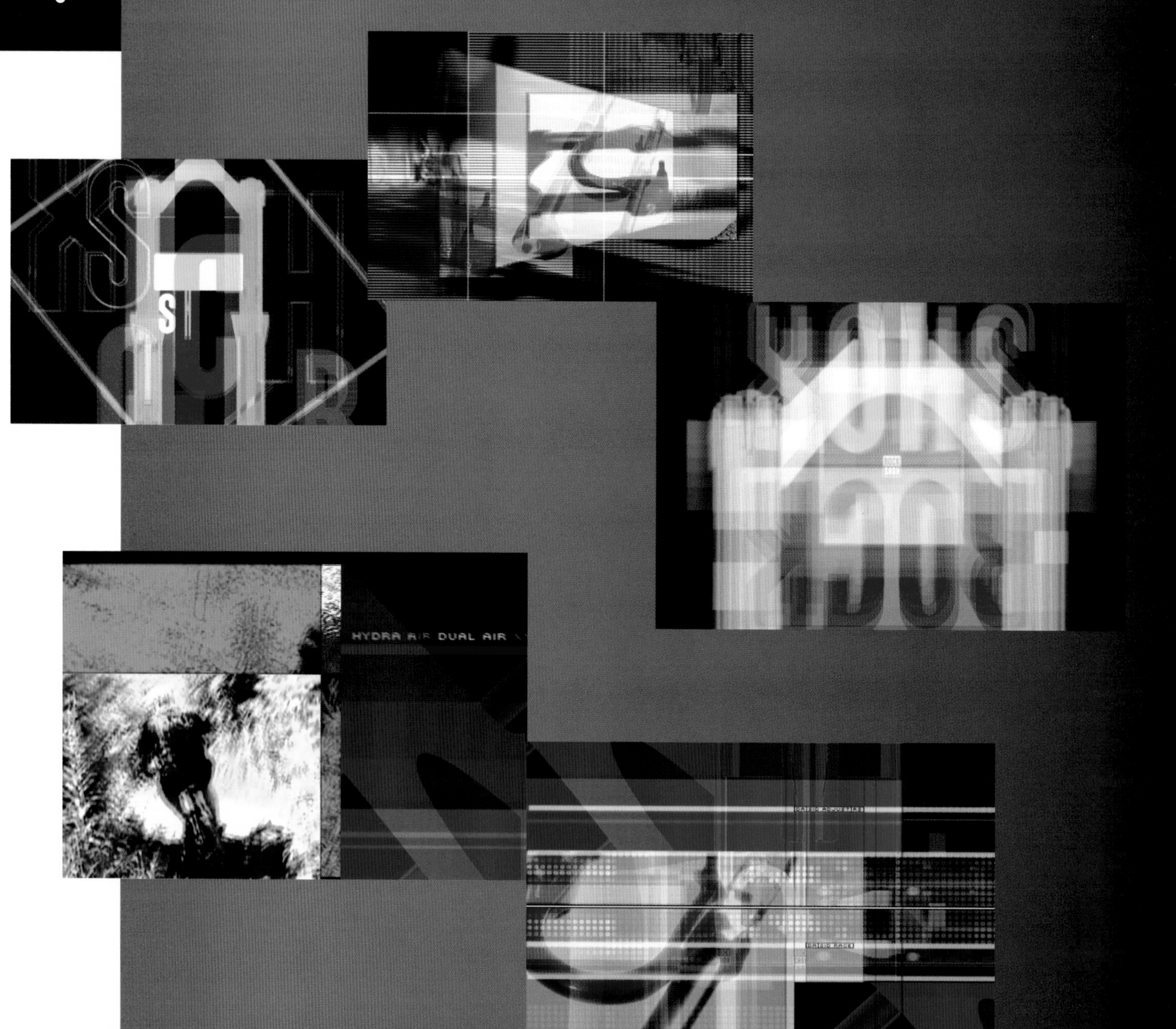

21

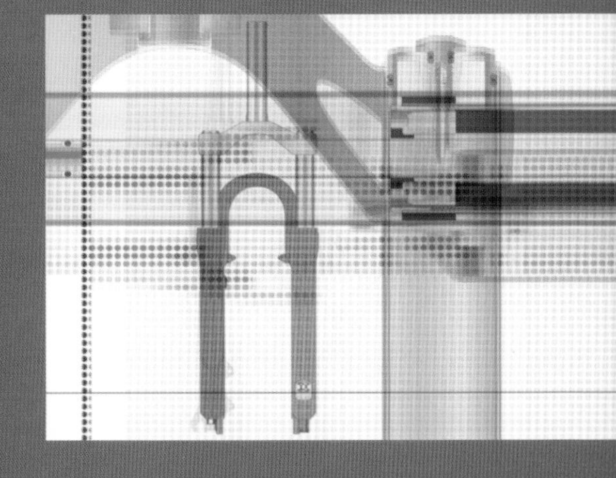

Show Opening **Motorama** Designers **Michal Oppitz, Tomas Moscak** Studio **Gratex International**

Motorama

Motorama is an auto-moto show aired on Slovak commercial TV Markiza, featuring the latest auto industry news, moto advice, reviews and reports from automotive sports. Gratex creatives Michal Oppitz and Tomas Moscak were charged with completely redesigning the visual style of the show including the logotype, show opening, jingles and graphic backgrounds. To entice a target audience of rabid moto-fans and conventional auto buffs, the opening jingle was energized using fast shots and many cuts. The celebrated design combines film footage with 3D graphics, typography and geometric elements.

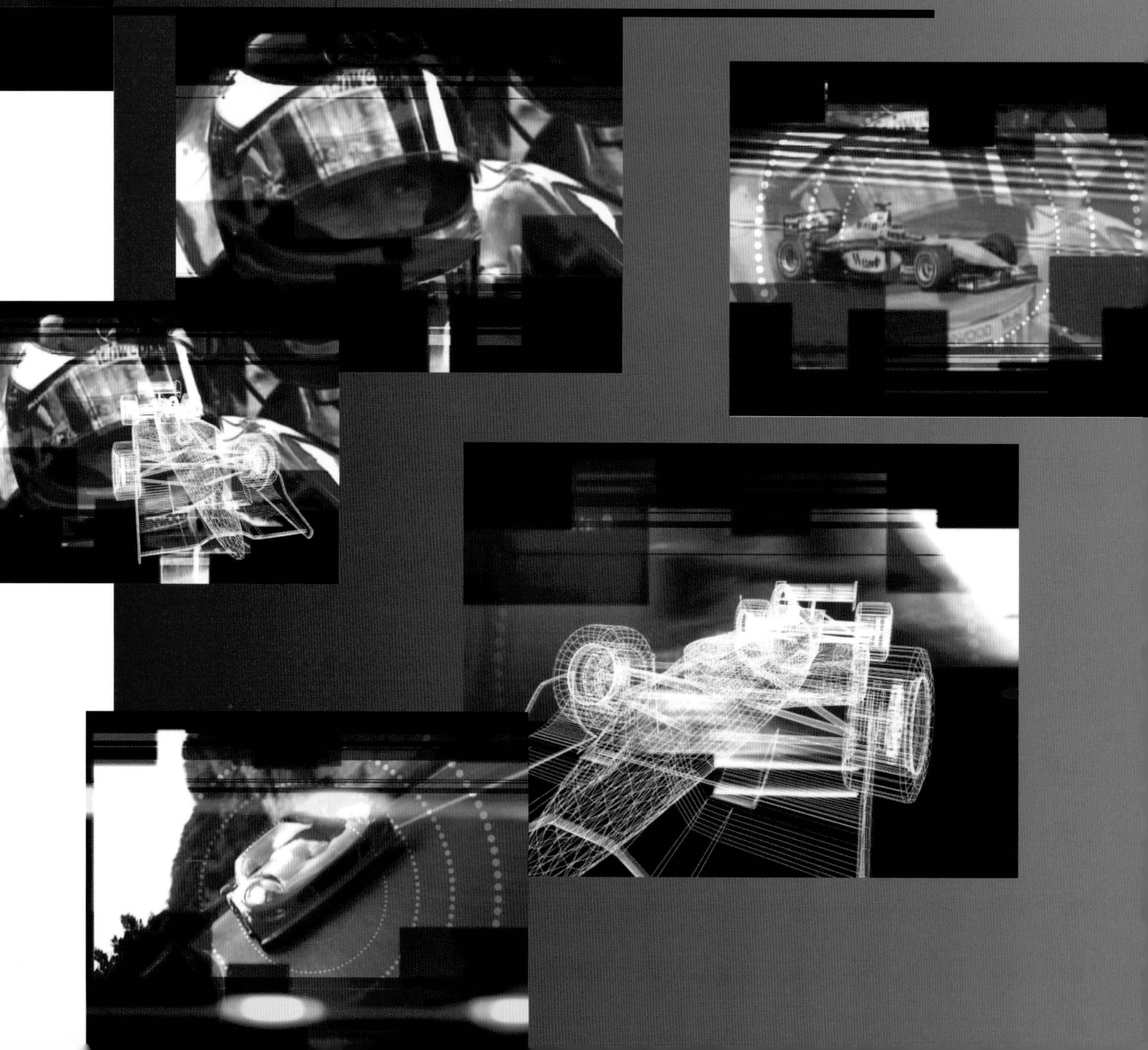

270
4 DEVIN
IVM

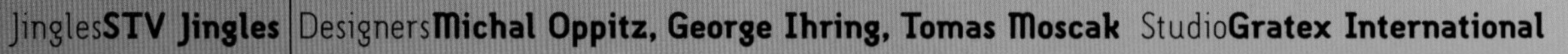

Jingles **STV Jingles** | Designers **Michal Oppitz, George Ihring, Tomas Moscak** Studio **Gratex International**

STV Jingles

Seeking a radical change, Slovak TV commissioned Gratex International to make it happen. STV wanted to distance itself from the Slovakian national colors of red, white and blue. The challenge was creating a new identity that appealed to a universal audience. The Gratex design team picked a fresh color scheme of yellow, orange and high-contrast blue for optimum impact onscreen. The new palette symbolizes the sunrise, a nod to the rebirth of STV as a progressive television station. Each promo is based on a single 3D object representing each theme, combined with typography and final effects. Featured are shots from the broadcast opening and theme jingles for programming ranging from children's television to evening movies.

víkend s stv

IPO Road Show Opening Video

To launch ZEFER into the financial community (before the Boston-based strategic Internet firm's planned IPO), Jason Zada and the in-house design group developed a 30-second video to precede the CEO's presentation. They created a tone of sophistication and innovation to showcase the company's unique culture, the fast-paced, visually stimulating environment cycles through colors, abstract imagery and ZEFER's people. The objective was to leave the audience on the edge of their seats before the presentation. With a team of eight creatives working around the clock, the final piece was shot, designed and edited in 48 hours.

Additional Design San Francisco ZEFER Design Teams Studio ZEFER Internal Design Team

Titling Sequence **Itch** | Designers **Kevin Wade, Abraham Animation** Studio **Planet Propaganda**

Itch

Planet Propaganda developed this infectious titling sequence for a television pilot focusing on unsigned basement bands and artists. Due to the diverse nature of the musical offerings highlighted within the half-hour show, the designers purposely created an "anti-music" audio track to avoid favoring any one category. The audio itself is itchy, building a frenetic foundation for the animation in the sequence. Scratchy line art and type fly around the screen like buzzing insects above a rash of soft-focused color. According to the Planet team, "It's all about scratching your creative itch."

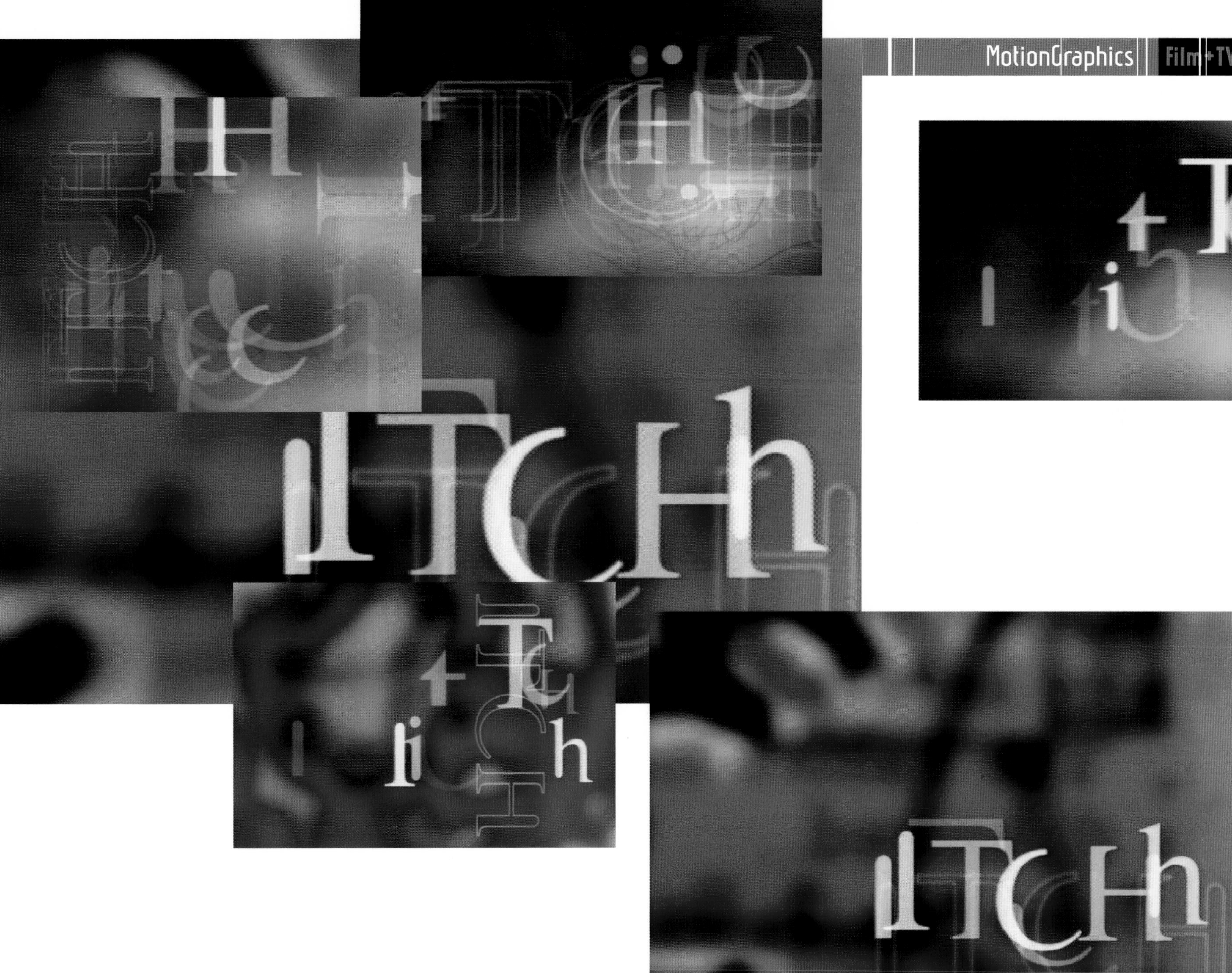

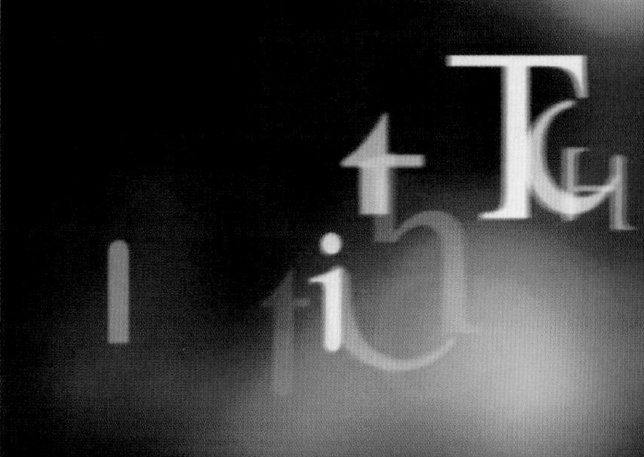

Opening Titles **AICP Midwest 2000 Show Open** | Creative Director **Joseph Kiely** Art Directors **Mike Czernuik, Joon Hong Park**

AICP Midwest 2000 Show Open

The AICP 2000 Show Open merges commercial industry logos with the transient life of its target audience in a frenzied rush-hour environment. Chicago-based FLUX Group's visual goal was to create a cohesive glue using color and inherent shapes present in an urban setting. The project's graphic climate was the antagonist of movement and interaction with the logos. The live action component, shot on super16mm, followed style boards with a guerrilla location scout and shoot. The transfer session was more traditional, adhering strictly to a specific feel, tone and palette. A combination of Adobe After Effects, Photoshop and Illustrator elements were used for the offline creative editorial and finishing.

Designers **Dave Parsons, Chris Strong** Creative Editorial **Colin Carter, Brooks Ruyle** Studio **FLUX group**

WKLU

TV Commercial **WKLU** | Director **Colin Carter** Creative Director **Joseph Kiely** Art Directors **Joon Hong Park, Sandy Weber**

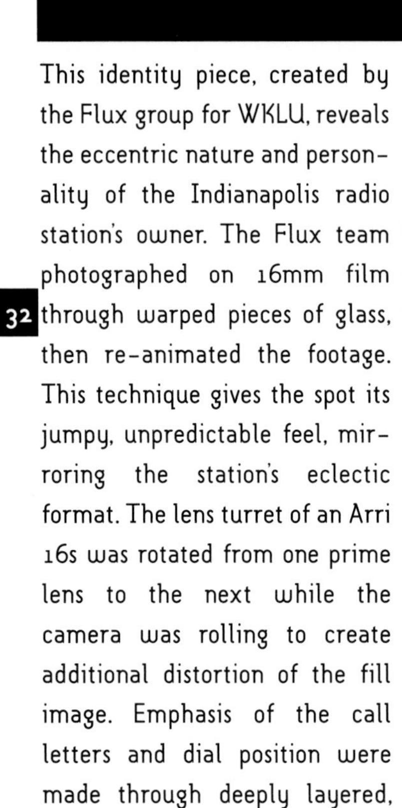

This identity piece, created by the Flux group for WKLU, reveals the eccentric nature and personality of the Indianapolis radio station's owner. The Flux team photographed on 16mm film through warped pieces of glass, then re-animated the footage. This technique gives the spot its jumpy, unpredictable feel, mirroring the station's eclectic format. The lens turret of an Arri 16s was rotated from one prime lens to the next while the camera was rolling to create additional distortion of the fill image. Emphasis of the call letters and dial position were made through deeply layered, kinetic title animations in Adobe After Effects.

Creative Editorial**Colin Carter, Brooks Ruyle** Studio**FLUX group**

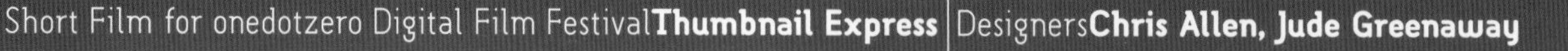

Thumbnail Express

Thumbnail Express is the opening chapter in the Light Surgeons' series titled "Gilligan's Travels." This audiovisual documentary is based on a recorded interview with Venice Beach street philosopher Robert Alan Weiser. Weiser's ramblings became the initial narrative anchor in this journey exploring the social and political state of America. The project combines DV footage, film, motion graphics and photography with sound design by Scanone. Thumbnail Express premiered at the onedotzero4 Digital Moving Image Festival in April 2000. The piece was commissioned by onedotzero, along with a live documentary mixed media performance, "Electronic Manoeuvres," in which the Light Surgeons fuse club culture with digital film making to create a unique, evolving "live" experience.

Sound Design **Jude Greenaway (AKA Scanone)** Studio **The Light Surgeons**

Video Installation**Faces** Executive Producer**Michael Davis** Supervising Producer**Brian Dressel** Designer**Jesse Jacobs**

Faces

"Faces" explores the aspects of seeing and meeting people every day. Producer Brian Dressel asks, "what do we remember about each face we meet? Do the features become interchangeable in our memory?" The design of this sequence includes textual treatments that echo human facial features. The images were shot on BetaSP outside LA nightclubs and edited on the Avid. The production team designed the sequence in Adobe After Effects. The piece concludes by showing various features—hair, eyes, noses and chins—as interchangeable. Dressel's project argues that faces do blend in our memory.

StudioSignCast, Inc.

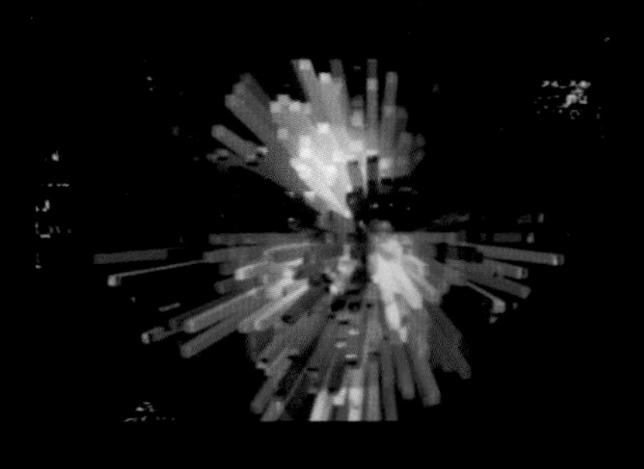

FACES

FACES WILL BLEND

Brian Dressel Personal Artwork

Video Installation **Brian Dressel Personal Artwork** | Designer **Brian Dressel** Studio **Brian Dressel**

Brian Dressel used kinetic images of schematics, circuits, liquid light and alien children to reflect modern technological society, de-evolved. The infinite zoom effect is utilized throughout the images, with the camera appearing to float through worlds of glowing circuits. The project combines still shots of schematics with live images captured on 35mm motion film in Chicago nightclubs. Zooms and video processing effects were created in Adobe After Effects. The sequence is accompanied by cold, dry, techno music, producing a feeling of distance and alienation from the humanity of old, and a non-specific concern about where it is all heading.

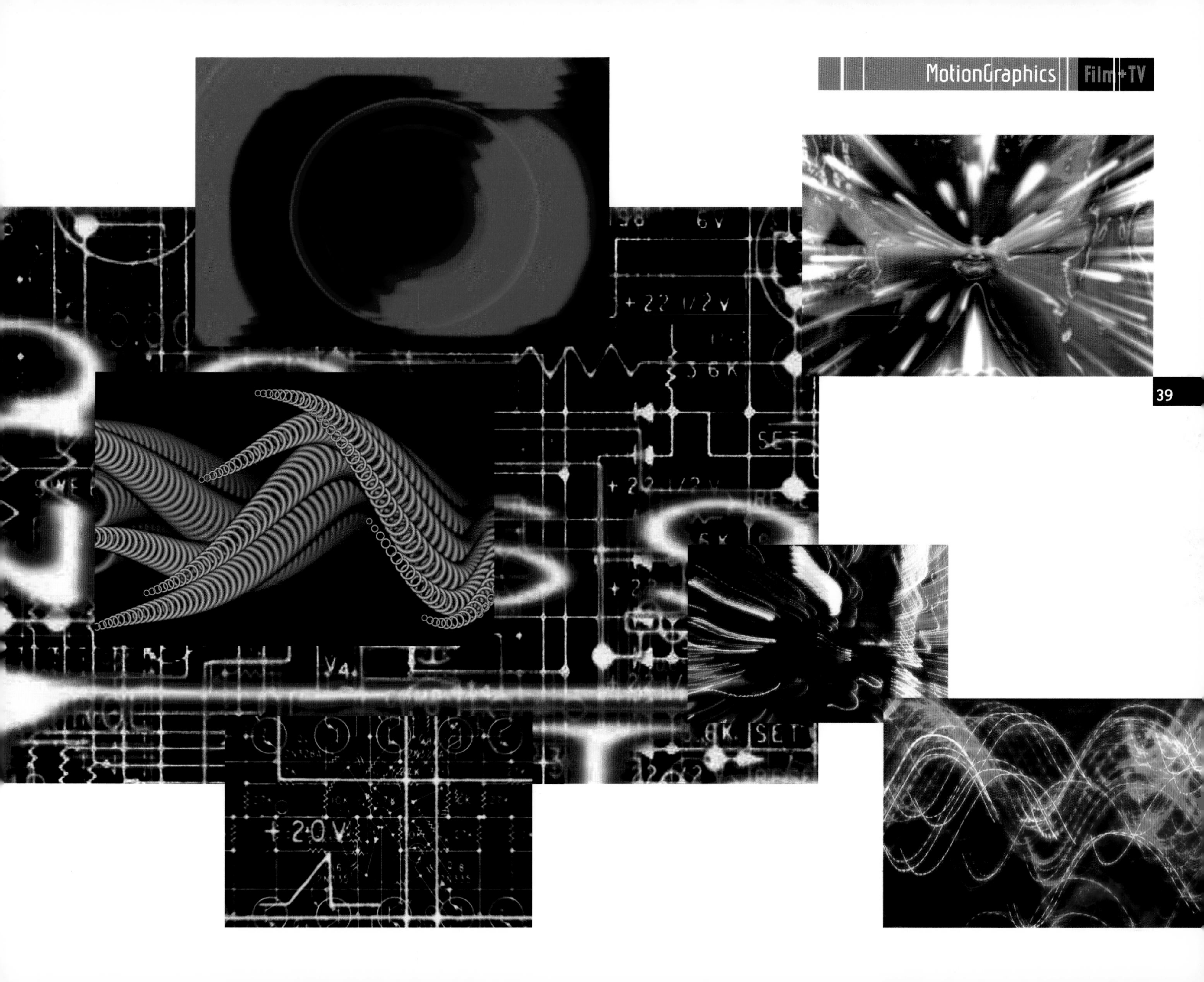
6V
+22 1/2 V
SET
V4
+20V

Cricket

Channel 9 Australia wanted a new look for their cricket coverage, one that would complement the new technology they were implementing this season. Engine delivered a design that layered statistical information over cricket players in action. The opener follows the basic story of bowler versus batter and the decisions made to win the game. Time-lapse clouds streak over a futuristic cricket ground. Spectators ripple with motion while statistic holograms appear above the crowd and playing field. 3D players on the field walk around as the day fades into night. Cricket was honored at the Australian Animation & Effects Festival for Best Title, Ident or Sting.

Title**Cricket** | Designer**Finnegan Spencer** Studio**Engine/Sydney**

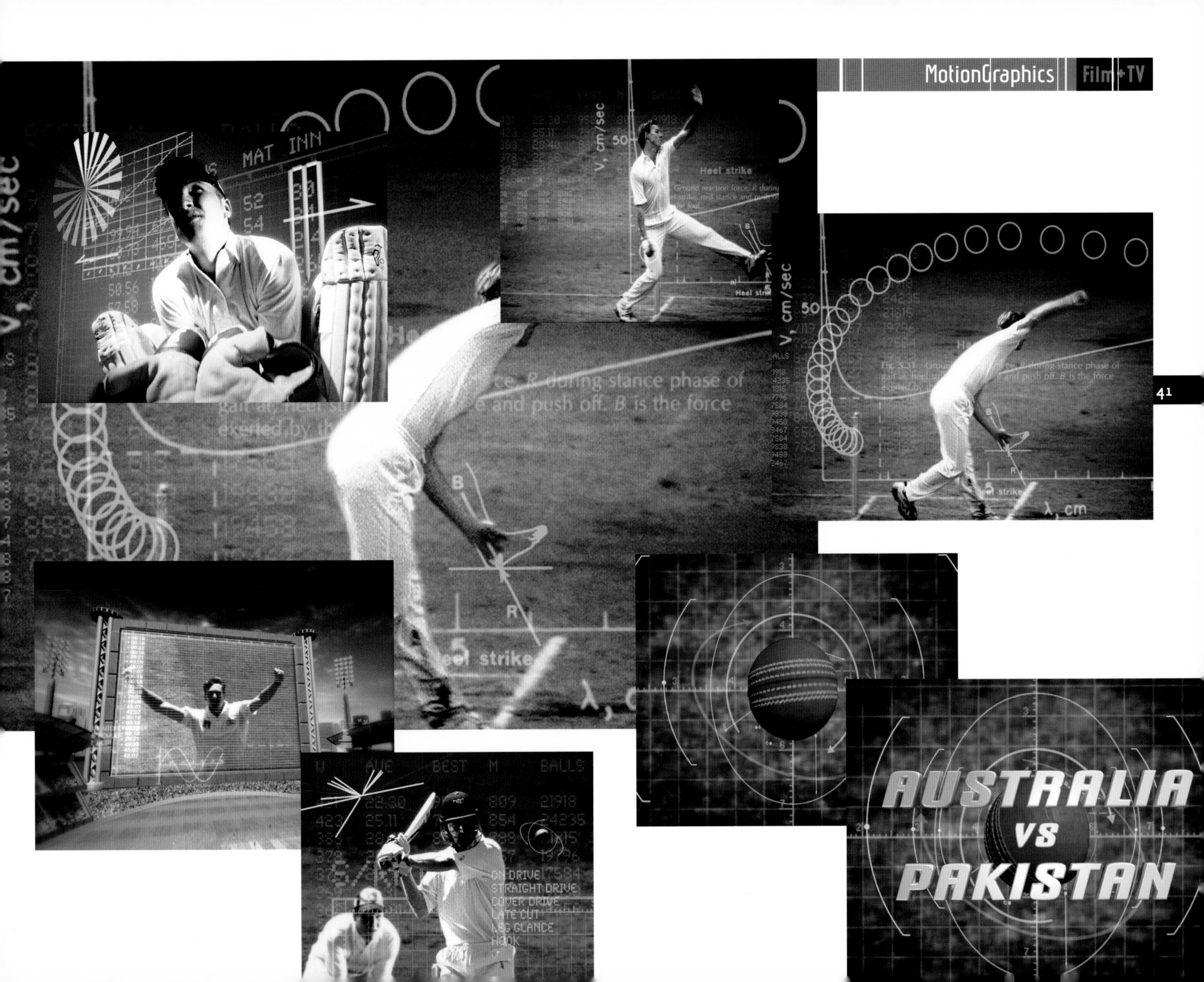

MAT INN
V, cm/sec
50
Heel strike
λ, cm
ON DRIVE
STRAIGHT DRIVE
COVER DRIVE
LATE CUT
LEG GLANCE
HOOK
AUSTRALIA
VS
PAKISTAN

Opening Title **Going Home** | Designer **Domenico Bartolo** Studio **Engine/Sydney**

Going Home

Going Home is an opening title sequence created by Engine for McElroy Television. The piece echoes the frantic vibes of the evening commute. Images of trains, stations and bustling crowds mix with fast-moving typography and graphics to create a visual rush. Shot on 16mm, the project was a journey of discovery not only in concept, but also in execution. The client was informed there would be no regular approval process—instead they had studio access at any time. This was a direct response to the TV series concept, which included shooting, editing and screening to air daily—a continual work in progress. This award-winning show opening mirrored the same concept.

:COOK DINNER
:RING MUM
:GO
102
Kilometres
T4035
:SADNESS
:STOP
:QUESTIONS
:HEADACHE
:FREEDOM
:ACCIDENT
:SENS.
NEXT STATION
:UNDERSTANDING
18:36:08
GOING HOME
OING O E

Station Ident **Fox Sports 2000** | Designer **Domenico Bartolo** Studio **Engine/Sydney Studios**

Fox Sports 2000

The future of sport; the dawn of a new millennium, the station ID by Engine revolves around the central icon of the neo-man. Tomorrow's champion, scanned, monitored, probed and tested—a phoenix awaiting flight. The project was shot on green screen in the studio with a Sony digital betacam. The Fox Sports logo was created in Maya and comped in Adobe After Effects. Typography and graphic elements were designed in Photoshop, with selected rushes imported into Adobe After Effects for grading and text animation. Scenes were imported into Avid for off-line editing. The process wavered in Adobe After Effects and Avid until all scenes were finished, with the project completed in Henry.

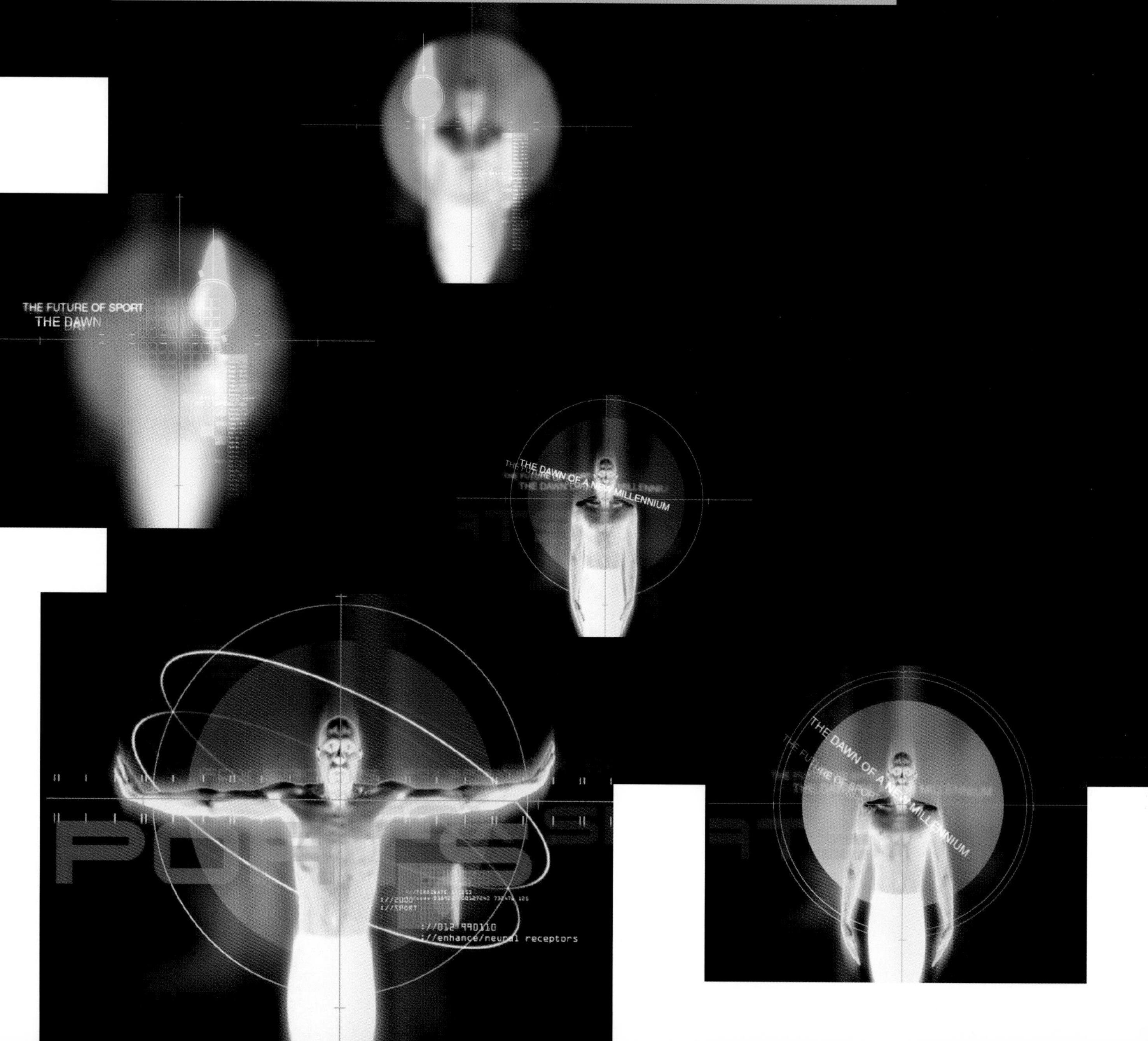

FOX
SPORTS

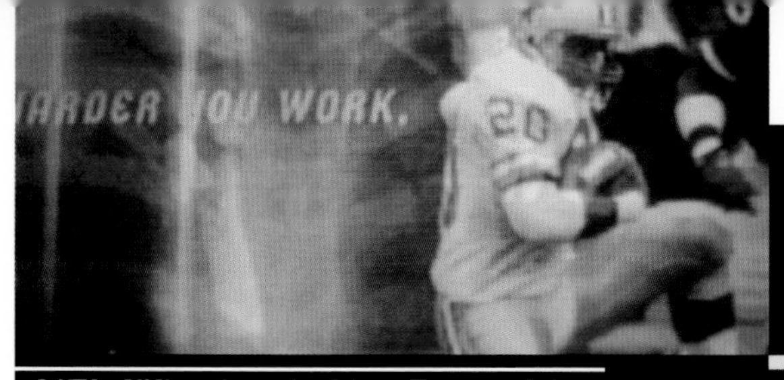

NFL Winning is Not Everything

"Winning is Not Everything," is a retrospective on NFL history created by the design team at Motion Theory. The audience relives the tense moments in the game and the exhilarating celebration of victory through the quotes of the immortal coach Vince Lombardi. Momentous plays are highlighted and framed through multiple layers and windows of graphic design. Kinetic visuals are held together through graphic elements reminiscent of football plays and technical manuals. Fractured typography builds tension and excitement with music leading to the chaotic energy symbolic of the action onscreen. The piece is dramatic yet fun, like a roller coaster ride through the history of the sport.

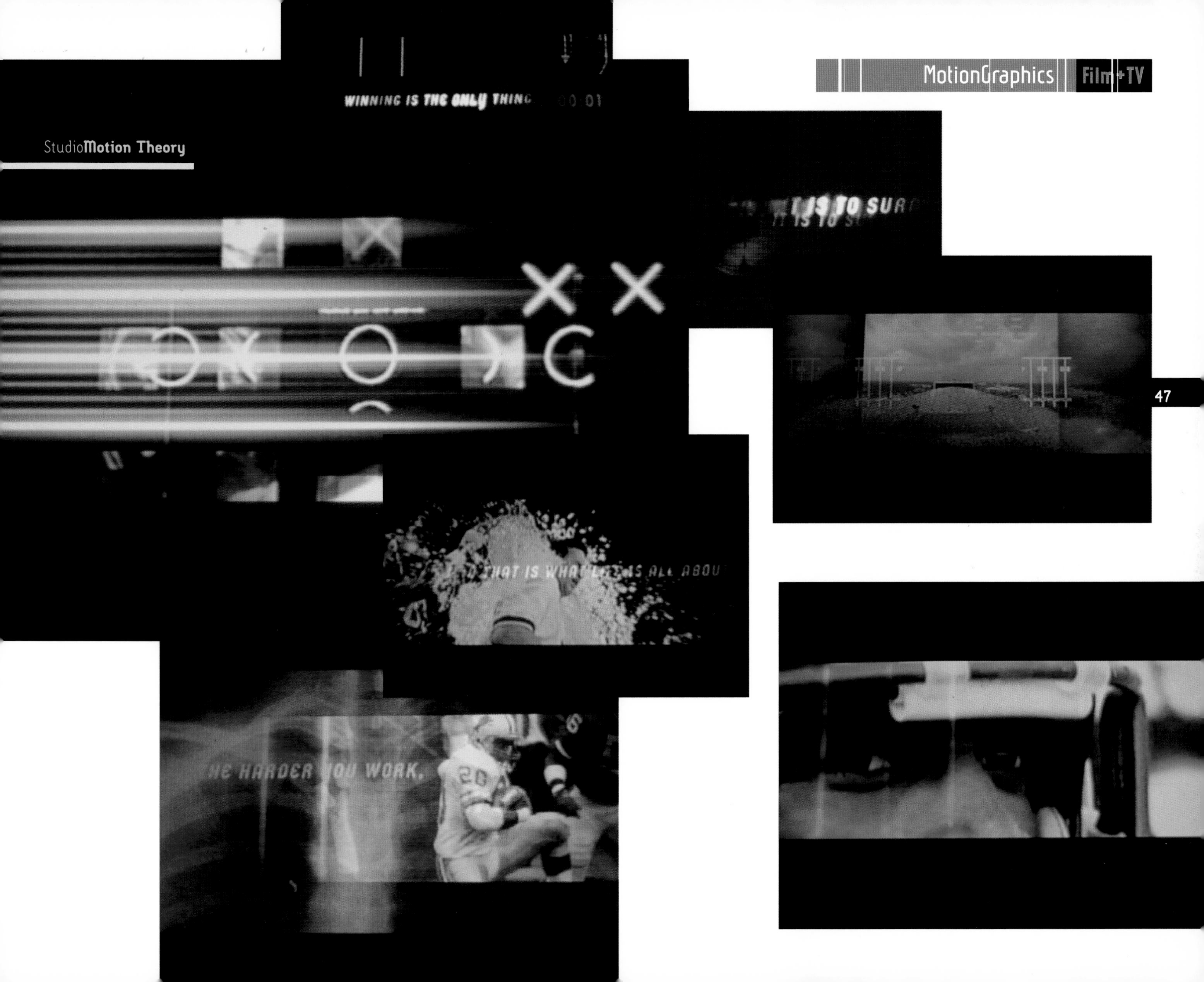
WINNING IS THE ONLY THING
00:01
THE HARDER YOU WORK,

Cinemax MaxTime Open

As Cinemax advances to a higher tier of movie premieres and a more sophisticated persona, Hatmaker and Cinemax continue their creative alliance. MaxTime, its primetime movie slot and first step in this transformation, stays true to the established brand, yet provides a lush, dimensional interpretation of the Max "dot." Video of particles moving through water was shot to create an organic quality as the viewer travels into the world of Cinemax. Graphics reinforce the time change while contributing to the environment. The feel is slightly reminiscent of a darkened theater and the wondrous emergence of light from the projector lens.

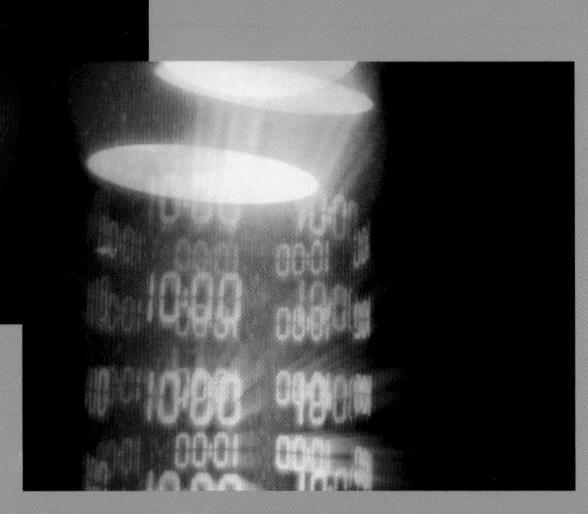

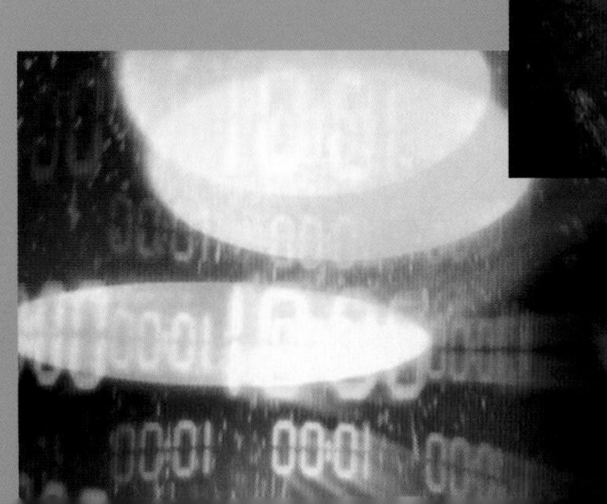

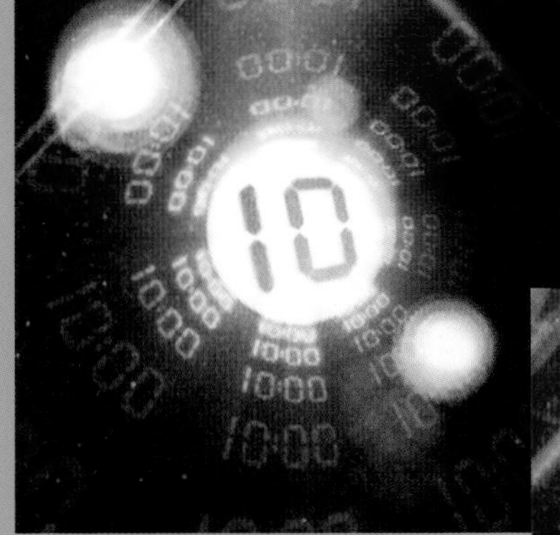
10
10:00

10:00

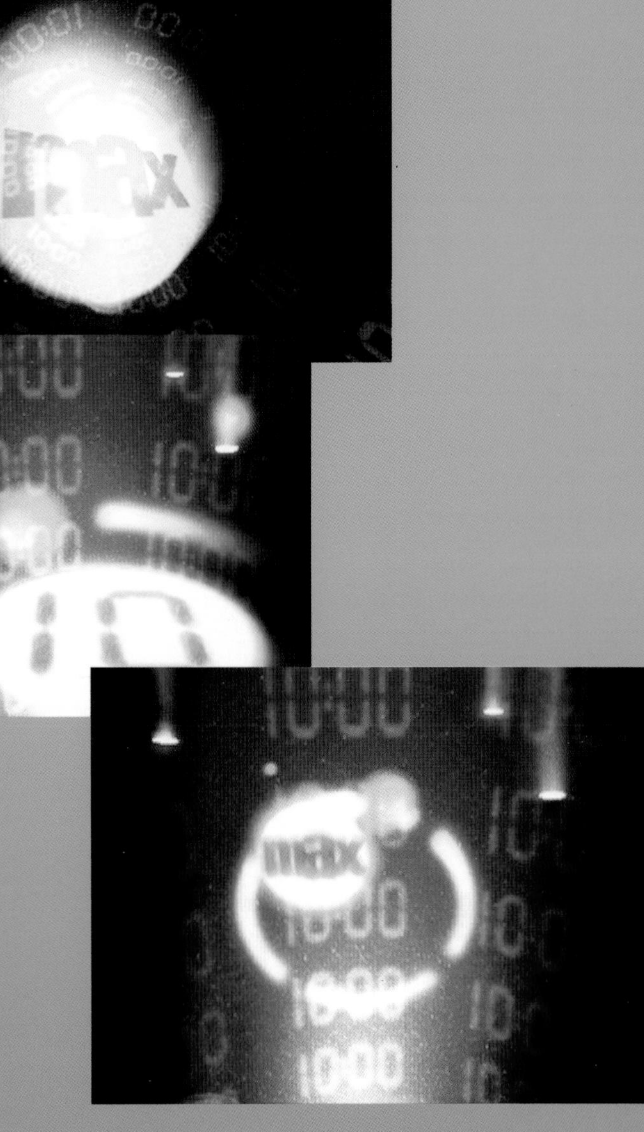
max
10:00
10:00
max
10:00

max
time

CourtTV Image Spot

Network Image Campaign **CourtTV Image Spot** | Designers **Goveia, Tharawanich, Bentley** Studio **Hatmaker**

Court TV came to Hatmaker for a fresh perspective—a means of elevating the network image and reflecting the spectrum of its programming. The solution takes a human view of the complex stories, people and facts. The Hatmaker team created a campaign that emphasizes the full cycle of crime and justice by highlighting aspects the viewer rarely considers. Streaming graphics over original 35mm footage lend a sense of urgency, while day and evening palettes help reinforce the tagline, "Judgment Days—Sleepless Nights." Window mattes in the "Next Ons" allow creative opportunities for Court TV's internal design staff.

81483549
RPD RSP
trial
trials
sleepless
nights
just
justice
judgment
days
COURT TV
judgment days
sleepless nights

CourtTV Movie Open

Primarily known for its reality programming, Court TV also has an extensive library of popular fiction and true story-based movies. The network needed a direct yet engaging way to relay this information to its audience. Hatmaker combined fresh takes on iconic cinematic imagery and suggested story snippets of original 35mm footage. A rich theatrical palette created graphics which embrace the genre, yet convey a sense of sheer entertainment. The production included Hatmaker filming the logo and putting the footage through a movieola for an effect reminiscent of true moviemaking and film.

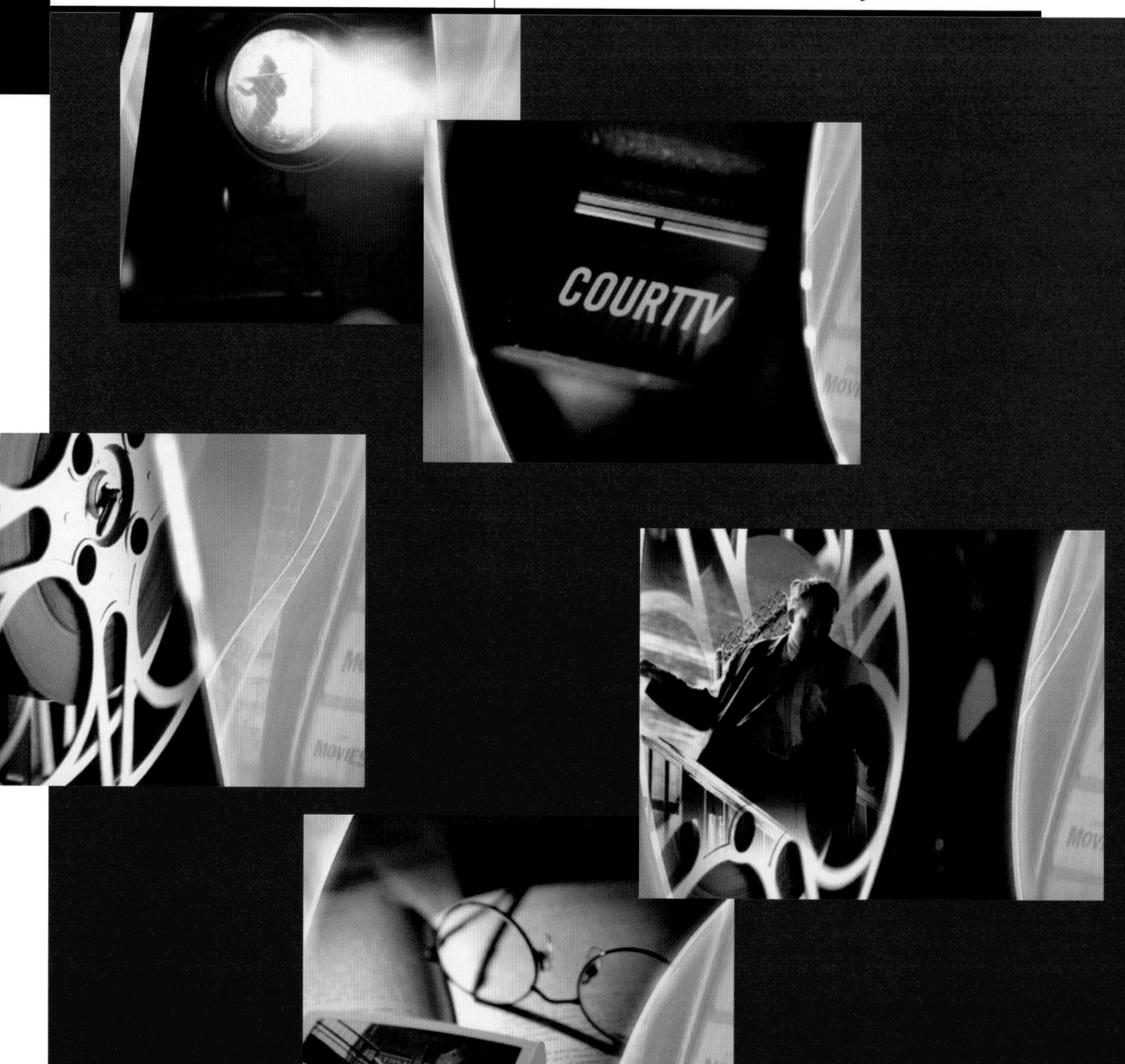

COURT TV
MOVIES

Oxygen Network Identity

Oxygen's groundbreaking "vapor" logo provides a flexible identity, reflective of the burgeoning spirit of the women's network. Building on the quirky, flat, graphic look of the network, Hatmaker (which designed Oxygen's logo) created "wallpaper" IDs—engaging patterns of graphic reveals set to originally scored, upbeat music. The wallpaper IDs are much like oxygen itself—sometimes overlooked, but always there.

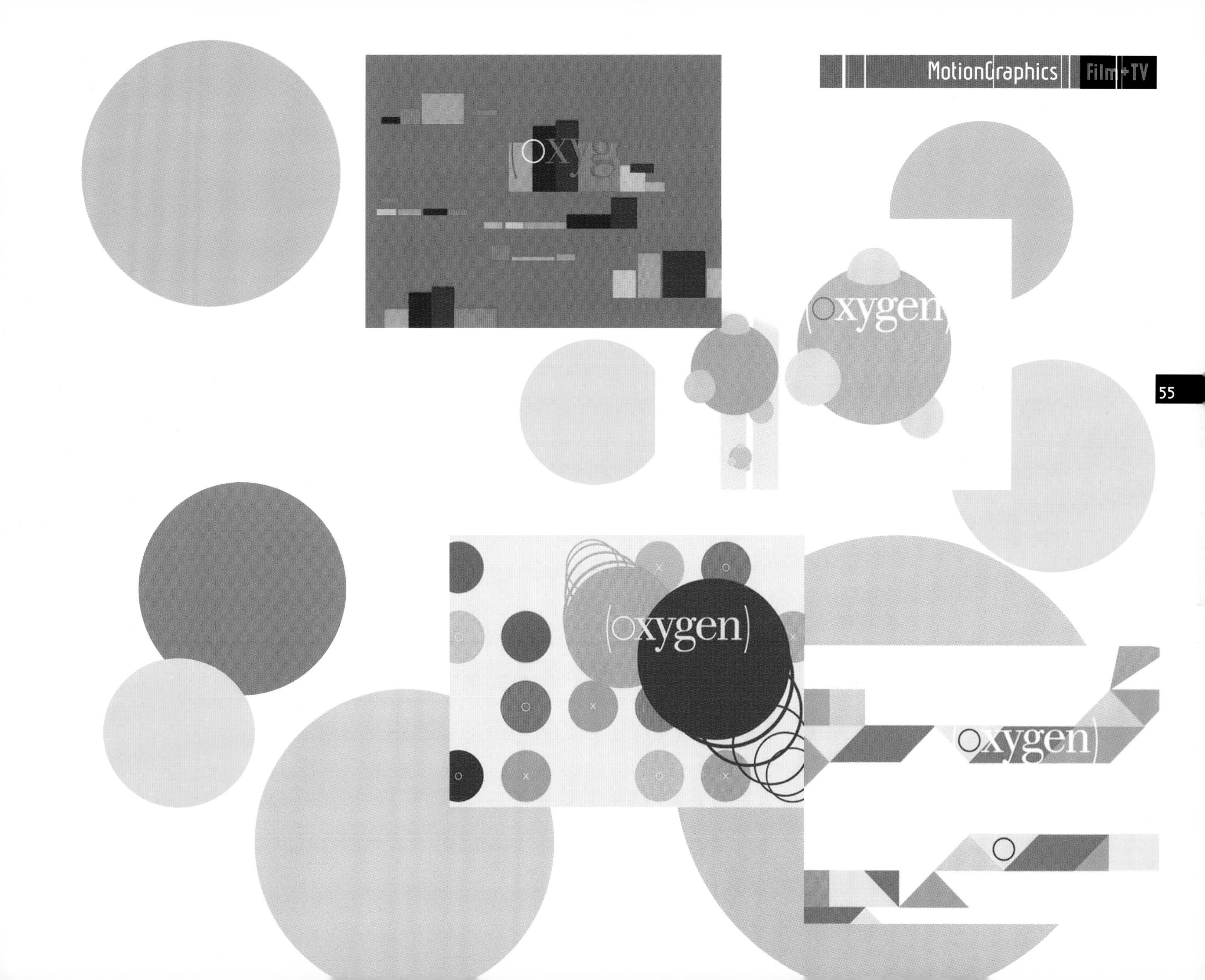
(oxygen)
(oxygen)
(oxygen)

DLJ Direct

DLJ Direct commissioned Ultraviolet to help promote their online investing service. Planes converge to create an infinite—and apparently random—stream of letters representing the enormity and chaos of the Internet. Within this plethora of information, a cursor leads the viewer to concealed textual phrases extolling the virtues of DLJ. The musical soundtrack spot relies solely on its visual content to deliver the message. Due to the relatively simple geometry, UV devoted most of its 3D efforts to the creation of dynamic camera moves that portray the work on a vast scale.

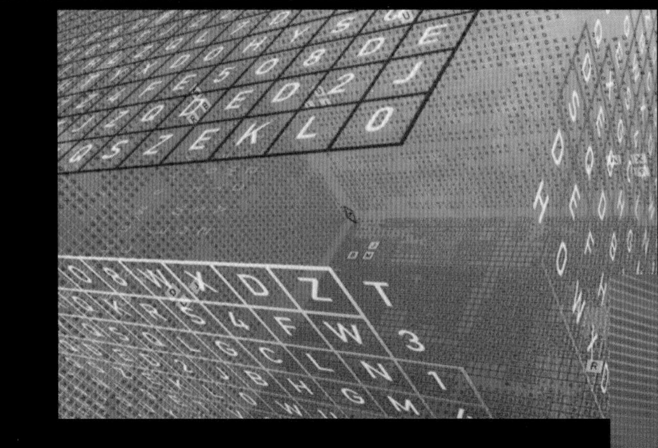

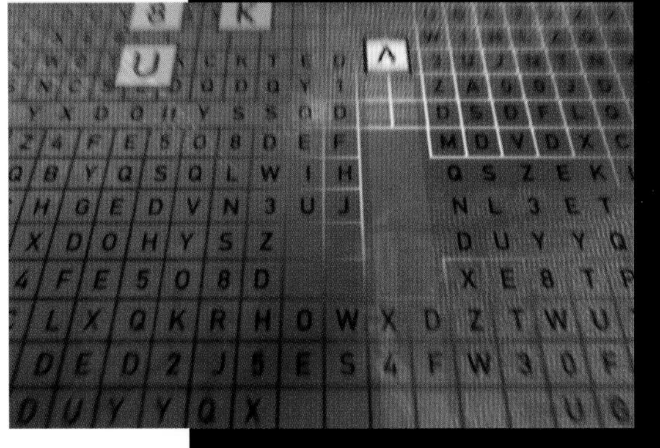

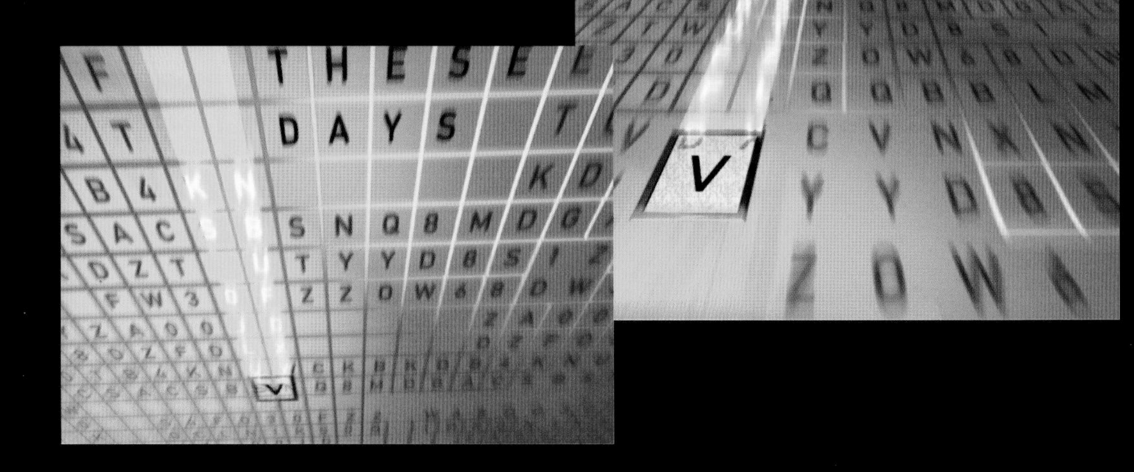

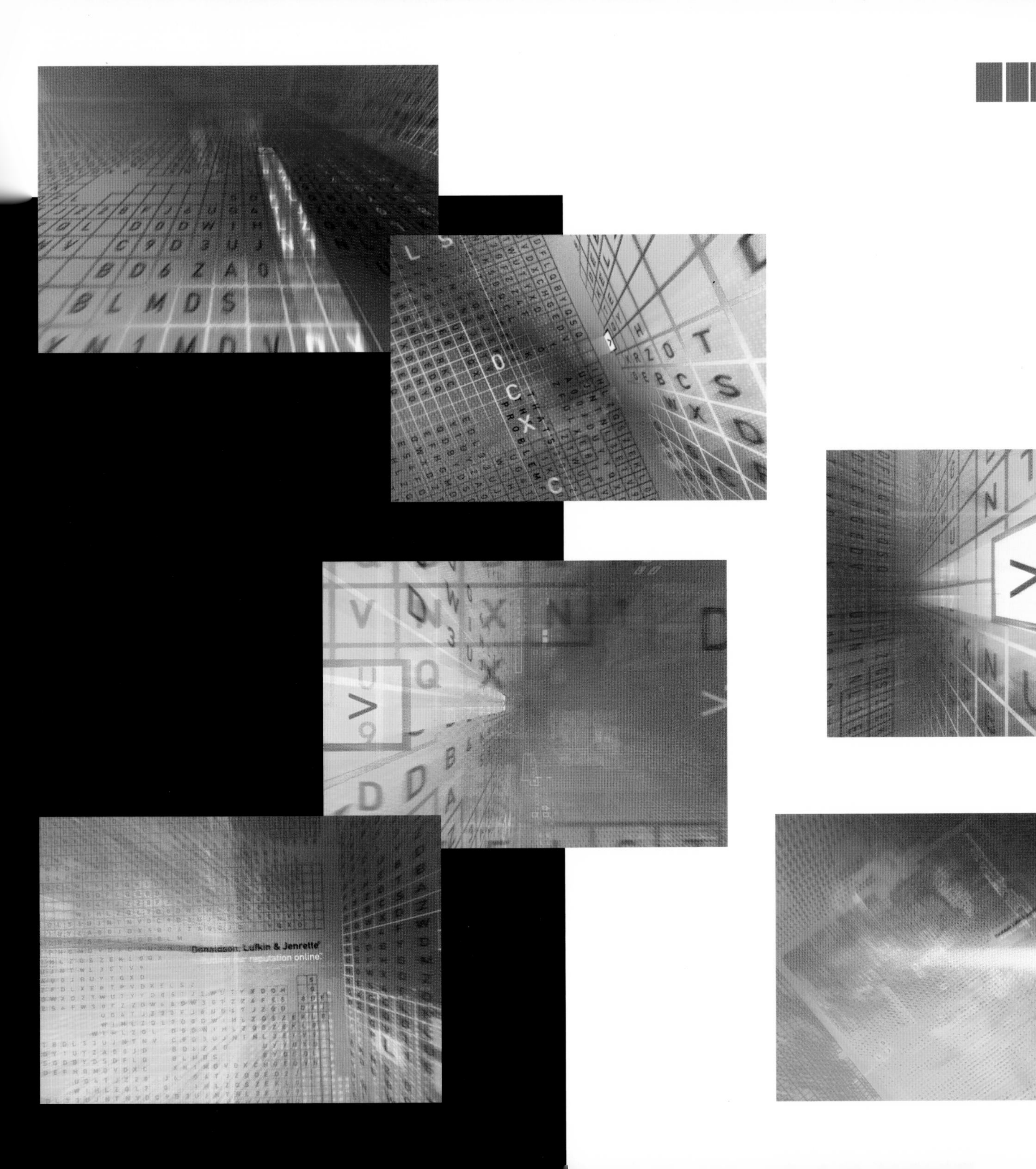
Lufkin & Jenrette
online.

Network Promo

Created by Ultraviolet to represent 25 years of Showtime, this :30 image spot visually explores an organic timeline made of light. The audience approaches the structure through drifting, epic camera moves, then dives into its core to discover a dynamic montage of Showtime's historical live action packaging. The project was designed primarily in Maya, with a heavy reliance on post processing for color and lighting. The volumetric lighting effects were created with the extensive use of radial blurs passed through cloud tank elements to lend an organic sensibility. Flame was used to further polish the visuals.

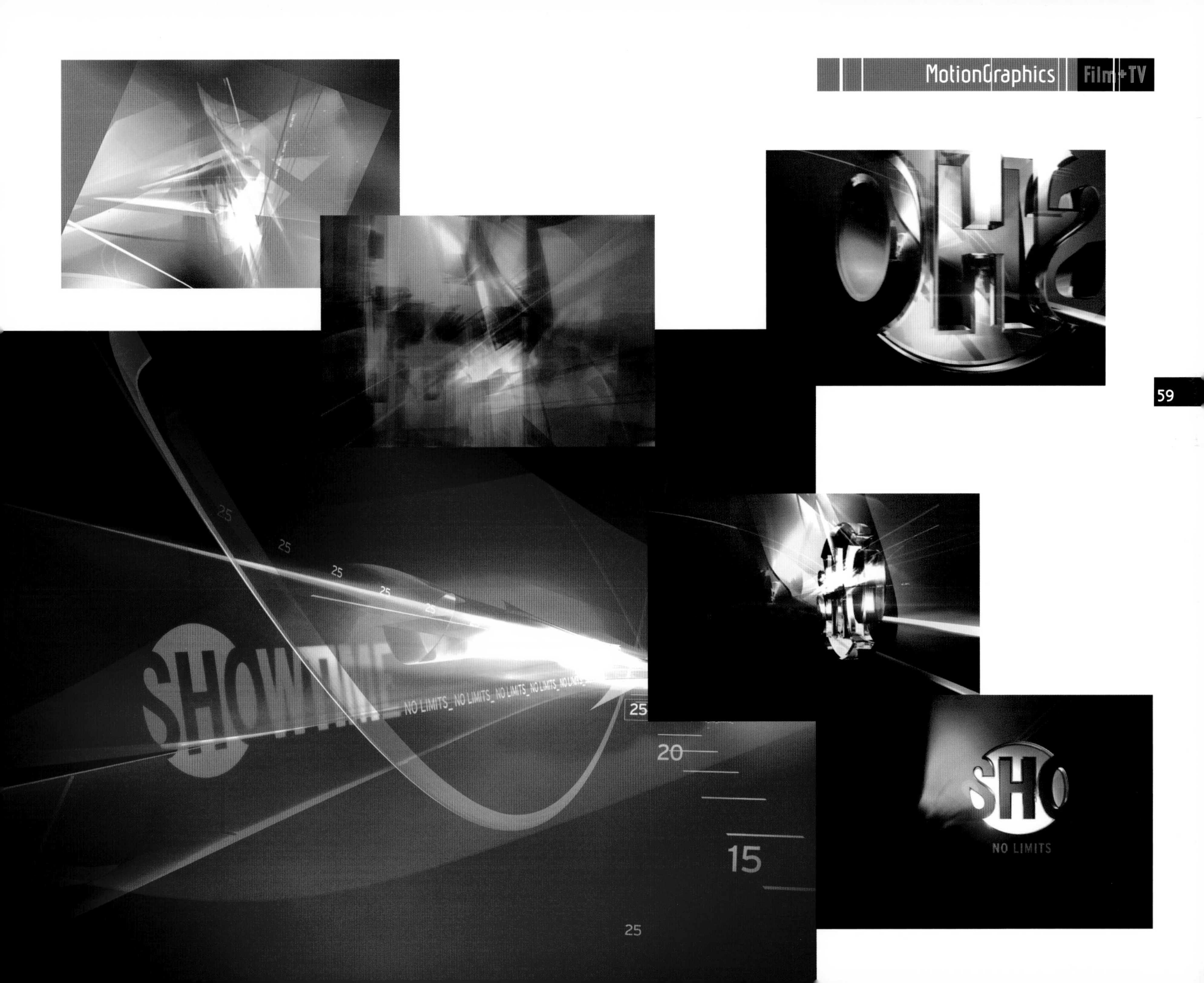
25
25
25
25
25
SHOWTIME
NO LIMITS_ NO LIMITS_ NO LIMITS_ NO LIMITS_
25
20
15
25
SHO
NO LIMITS

Movie Opening Sequence **VH1 Strange Frequency** | Designers **Jeff Linnell, Beat Baudenbacher** Studio **Ultraviolet**

VH1 Strange Frequency

Designed by Ultraviolet, VH1's "Strange Frequency" is described as the "Twilight Zone" with a heavy dose of rock 'n' roll. Suspenseful tales mix rock mythology with the unknown. The drama's titles are a journey through a world of iconographic moments. The sectioning of the imagery reflects the four dramatic storylines that structure the narrative. Using an oscilloscope and a vintage tuner, radio graphics are used as a visual thread throughout the opening, appearing subtly in many of the light sources and highlights. Smashing guitars, radio tuners, crime scenes, dark rockers, turntable rpm indicators and a disco ball are the hallucinatory rock 'n' roll-edged images that "Strange Frequency" invokes.

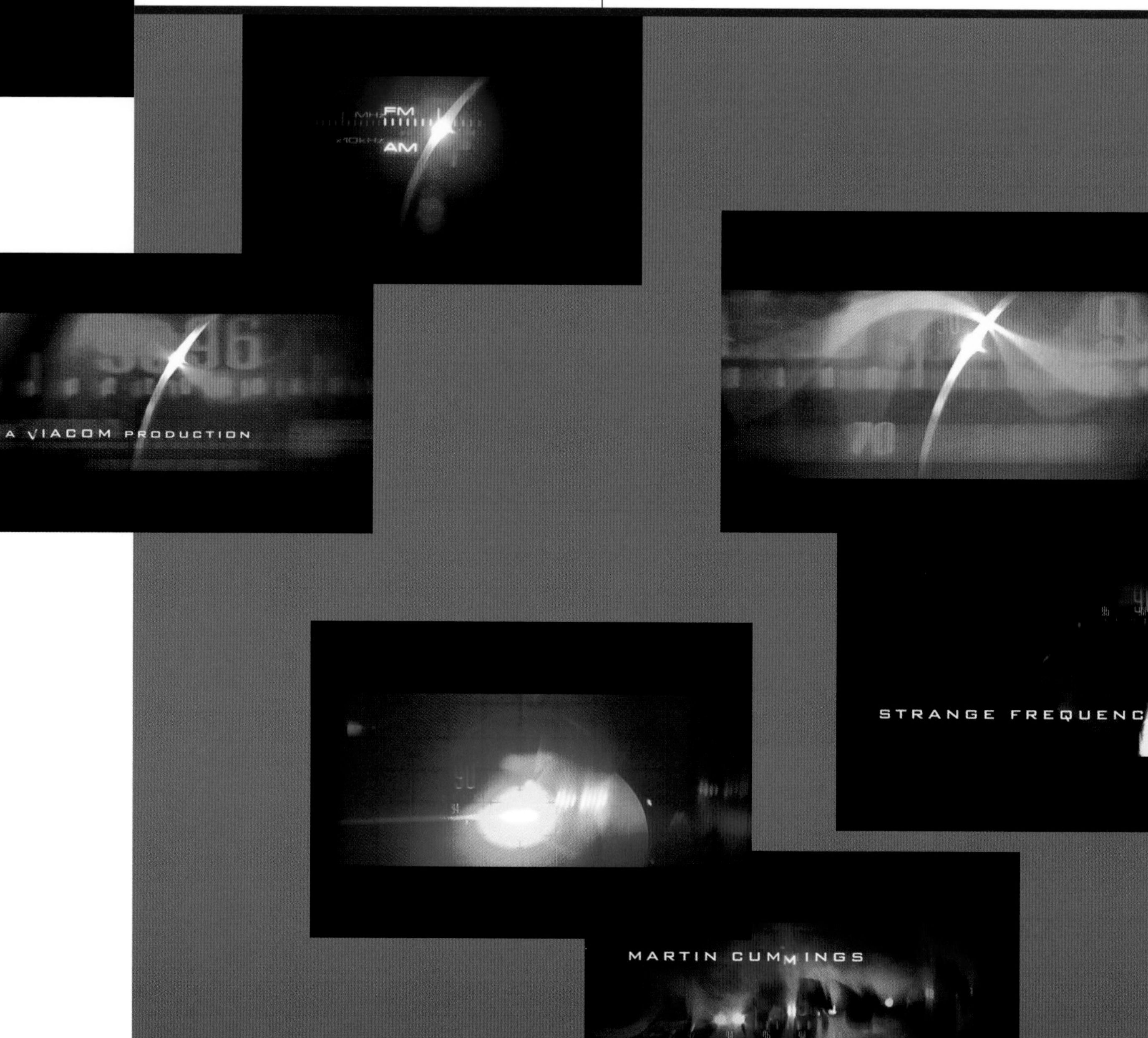

OKOLOFF
HOLLAND TAYLOR
JOHN TAYLOR
STRANGE FREQUENCY

VH1 Bands on The Run

"Bands on the Run" is a reality-based program that follows the travails of four independent bands competing on a 13-city tour. The Ultraviolet team chose a visual approach that would reflect the "lo-fi," do-it-yourself aesthetic of Indie musicians. The opening sequence depicts the onerous lifestyle of a rock 'n' roll tour. To signify the reality that the show embodied, UV chose photocopied event flyers as the main visual thread of the package. The UV team filmed the bands in Los Angeles and New York using a hand-held camera. The footage was copied on an old Mita DC-1205 to create the photocopied world that the flyers invoke.

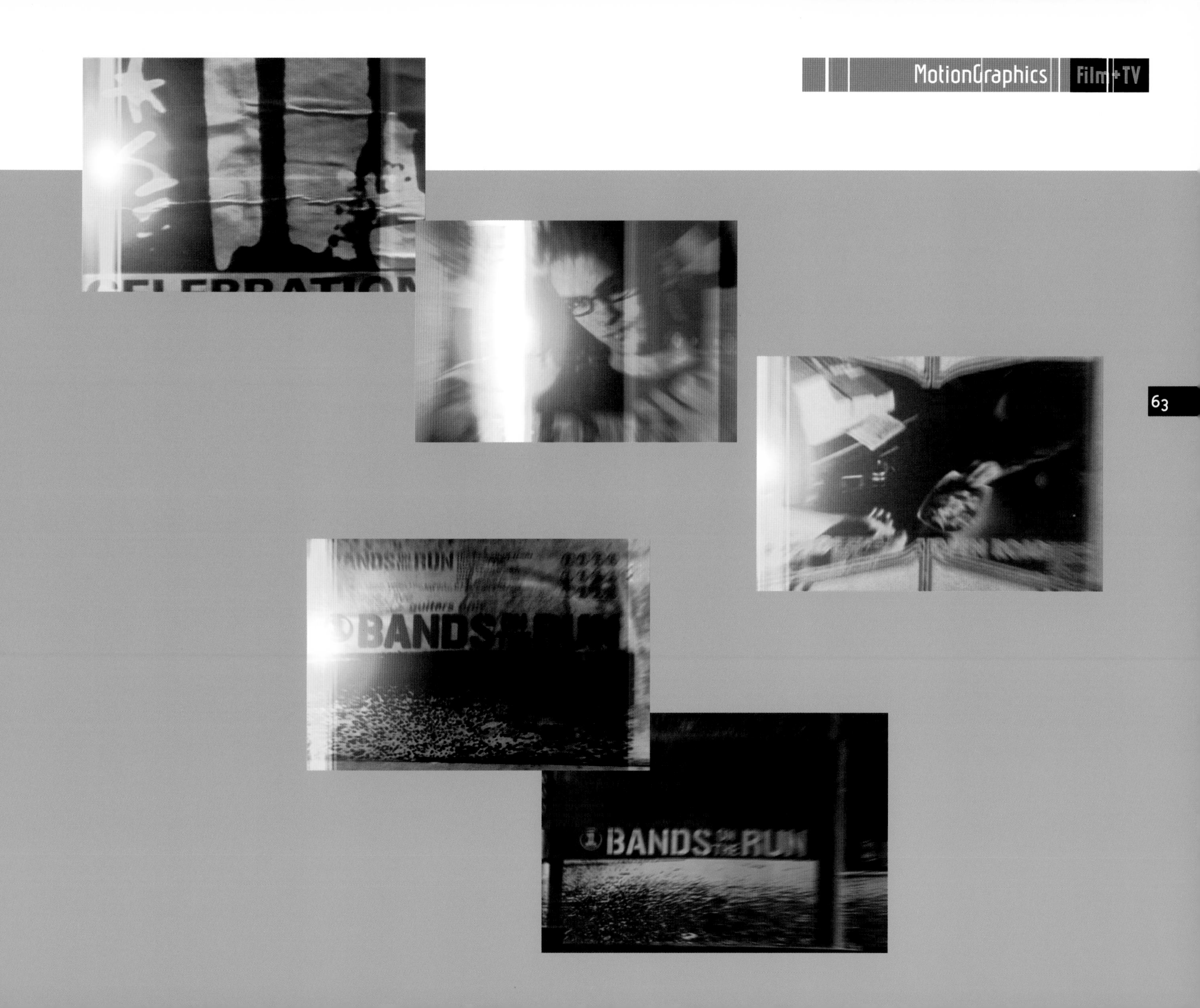
BANDS
BANDS

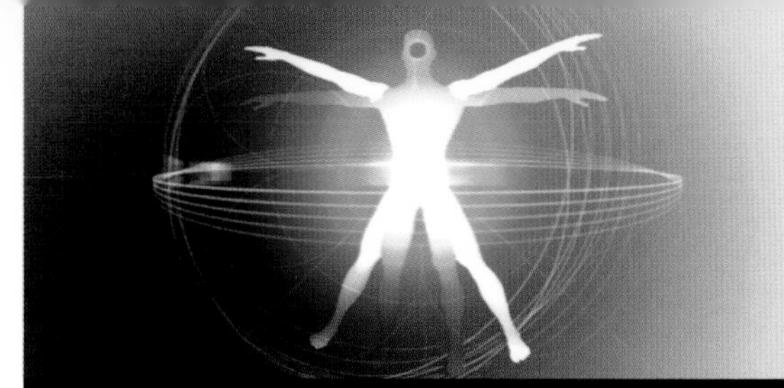

MTV 360 Transmission

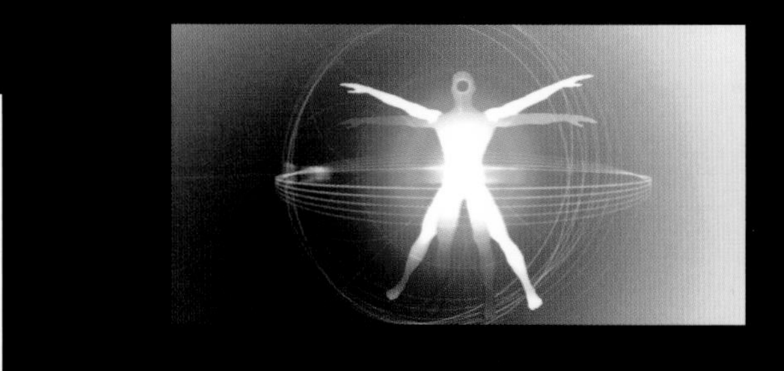

Ultraviolet, Phactory Productions and the Pixel Liberation Front collaborated to create a series of promos for MTV, "Circulation," "Transmission" and "Perception." Part of the MTV360 campaign, the promos heralded the channel's three media manifestations: MTV, MTV2 and mtv.com. The campaign presents the network within the microcosm of the human organism. A three-dimensional humanoid based on Leonardo da Vinci's Vitruvian Man visually correlates the three image spots. MTV's media forms are represented by the biological systems associated with the heart, spinal cord and brain. To close the spot, the Vitruvian man is reprised as the backdrop for the MTV360 tagline, "We revolve around you."

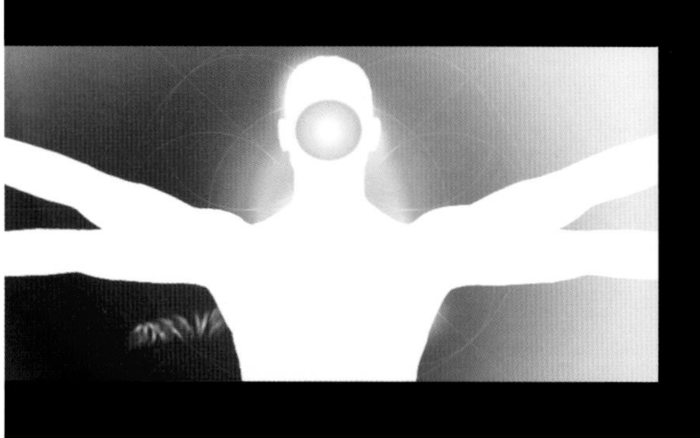

mtv.com
icon:christina aguilera

The Hives "Main Offender"

Stockholm's StyleWar was commissioned to create a new music video for Swedish Garage punk revivalists, The Hives. In this video for "Main Offender," the Hives fight the Negatives, who try to steal their position of power. The band was shot in front of a blue screen, and environments were 3D-generated parallel with editing and animation. StyleWar developed the piece in black-and-white to symbolize life in Punkrockcity—a cold, hard place where the sun doesn't shine all the time.

THE NEGATIVES
LEBOX
BONNEVILLE

Morning Glory

StyleWar's design team developed this fresh program identity for MTV Nordic. The concept behind "Morning Glory" was to recreate the sweet feeling of awakening spontaneously on a lazy Saturday morning. The piece was animated in 3D, with stark white graphic elements composited together with breathtaking images of the sunrise shot on 35mm. The show's ornate fraktur logo treatment lends a heavenly gothic edge to the serene backgrounds of the title sequence.

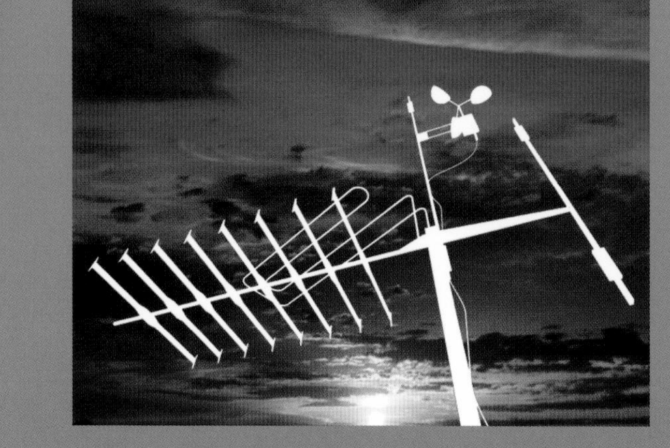

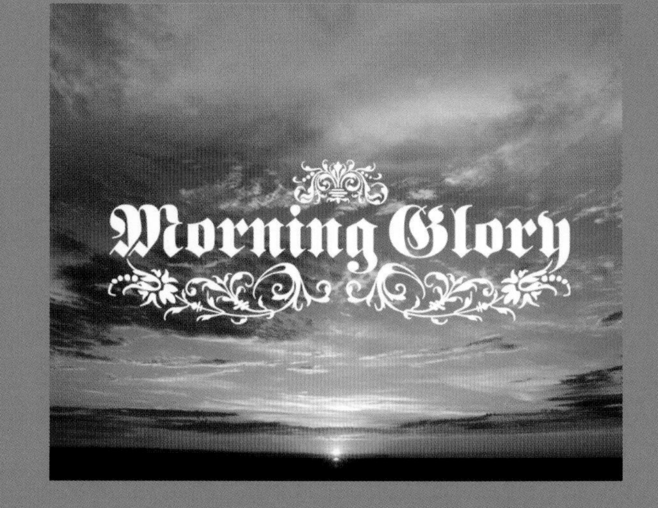

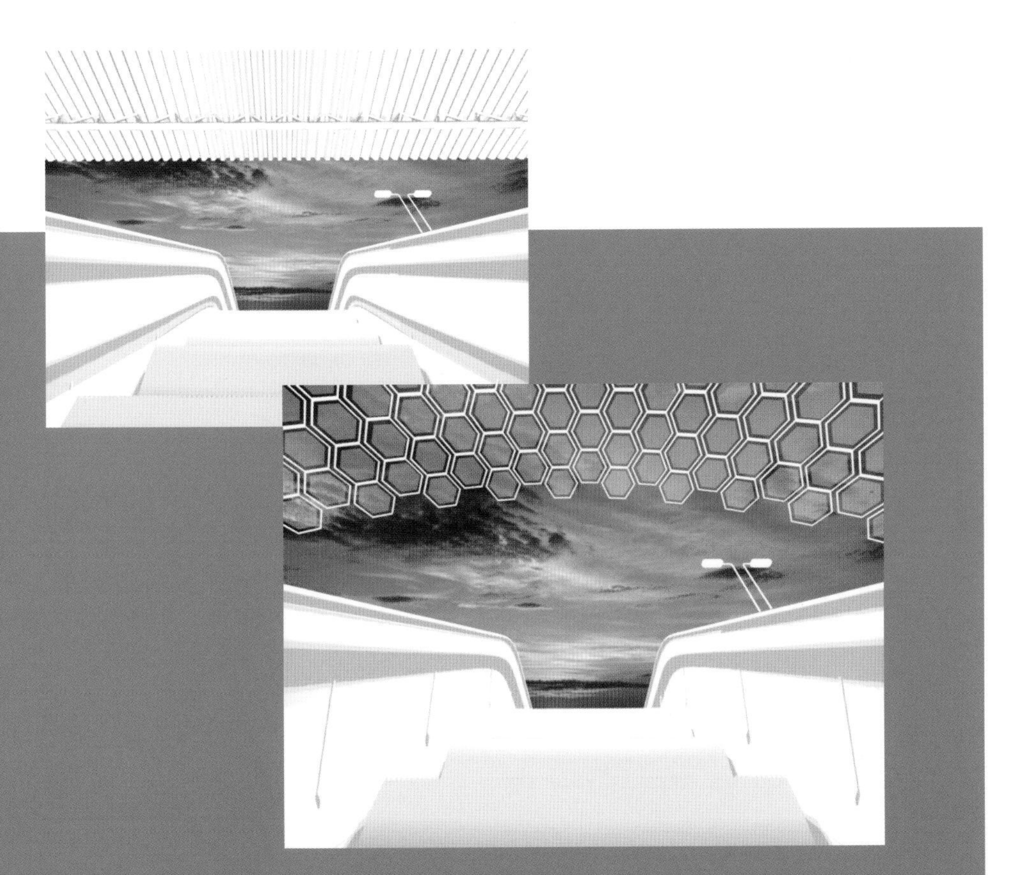

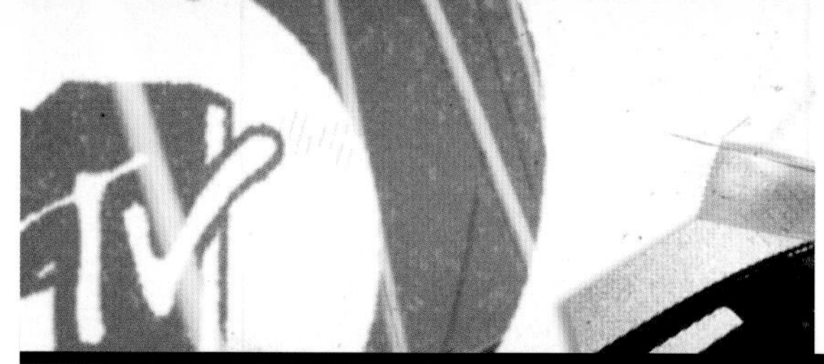

MTV-S Feria Promo

One of three IDs commissioned by MTV-S, this piece had to appeal to rock music fans. Unburro decided on a mixed-media approach, giving the project more of a filmic look, rather than a computer-generated appearance. The creative team shot D V footage at a day fair, then digitized it on the Macintosh. To achieve a roughed-up look, they scanned strips of clear tape and crumpled paper to add texture. To get the logo into the proper distressed condition, the Unburro designers faxed and refaxed a black-and-white version to each other, then scanned it for final tweaking in Adobe Photoshop. The elements were stitched together and animated in Adobe After Effects.

MTV

:30 Commercial **Ultimate TV** | Creative Director **Chris Do** Lead Designer **Tom Koh** Designer **Steve Pacheco**

Ultimate TV

Conceived as a journey from television's analog past to its digital future, this 30-second commercial playfully manipulates a few visual tropes of classic 1950's sci-fi shows before launching the viewer into the new world of digital video. BL:ND created this sequence using a combination of live action film footage, Beta videotape and computer-generated elements. Polished and refined in Adobe After Effects, the individual elements were then composited together on a Quantel Henry, and underwent further refinement in color, tone and the many subtle details that enrich the spot.

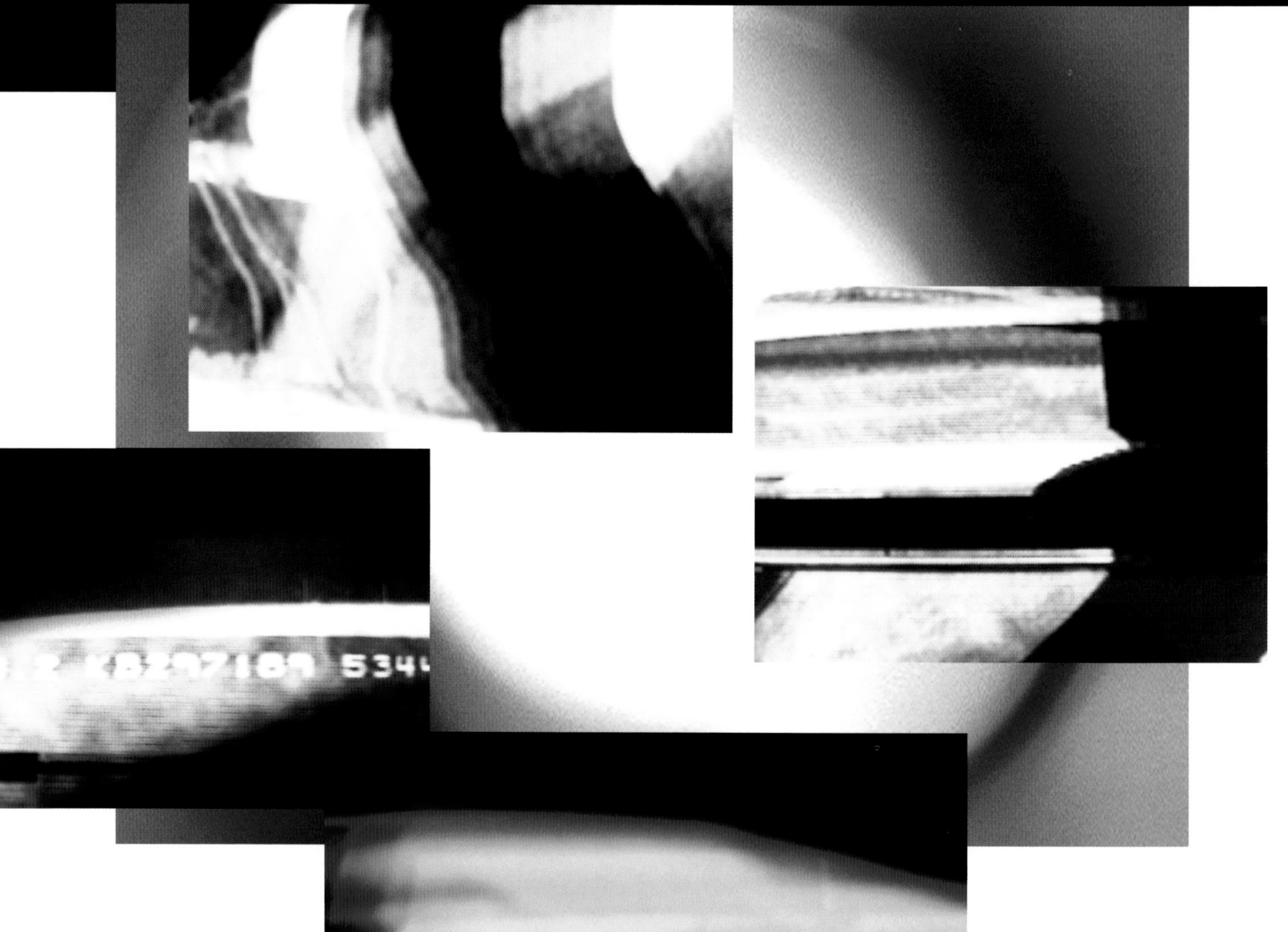

Animator**Lawrence Wyatt** FX Photographer**Rick Spitznass** Studio**BL:ND**

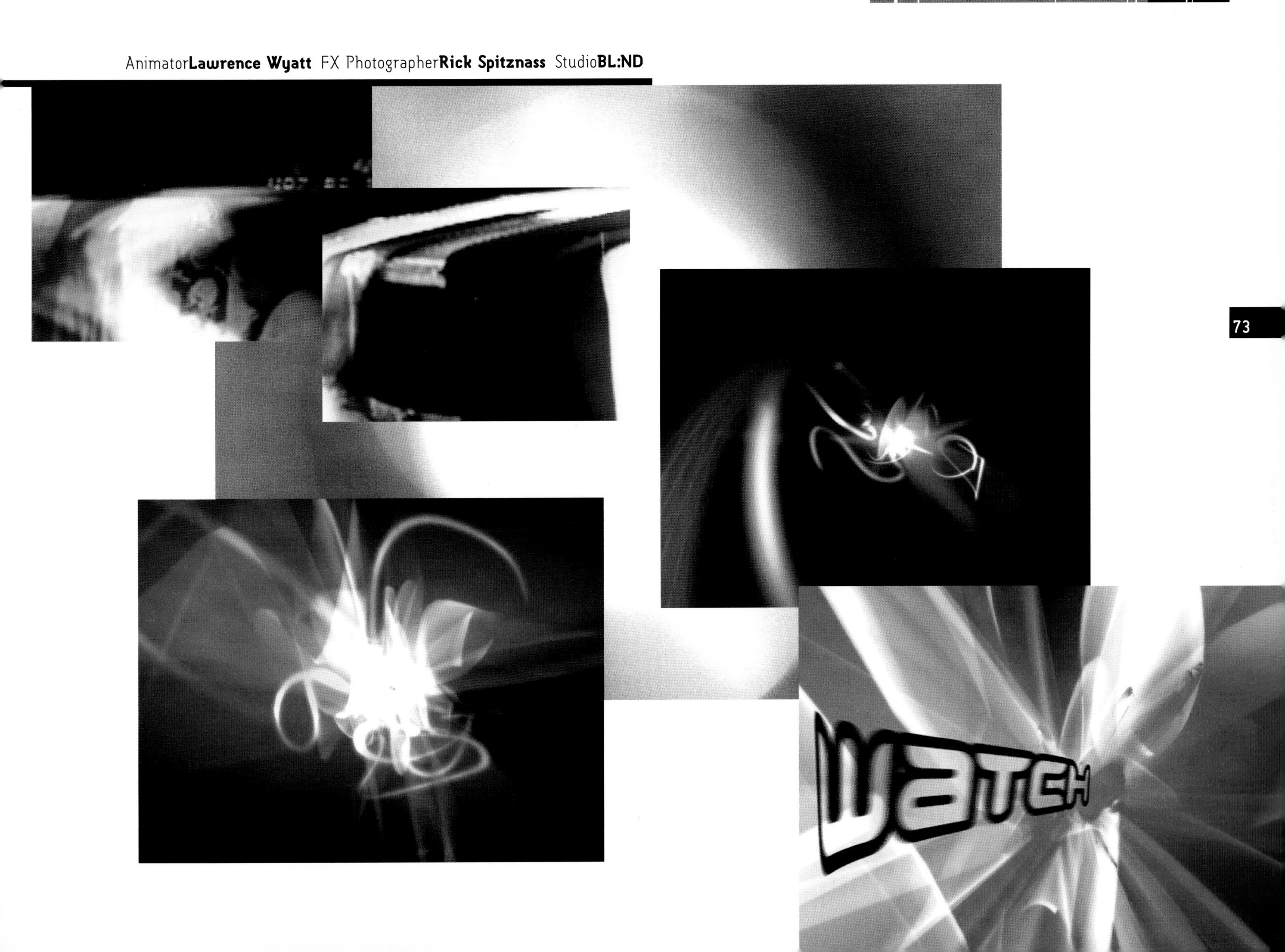

Dogtown and Z-Boys

Directed and co-written by skateboard legend-turned-filmmaker Stacy Peralta, this Sundance Award-winning documentary chronicles the exploits of the legendary cadre of young Southern California skaters known as the "Z-Boys." In the mid-1970's, their gravity-defying stunts and aggressive attitudes revolutionized skateboarding, and laid the groundwork for the emergent extreme sports movement. Using both Avid and Adobe After Effects, the BL:ND design team fused a variety of complementary elements, including stock photography, film textures and skating footage to support and enhance the mixed media effects throughout the film.

Animators **Calvin Lo, Wilson Wu and David Ko** Studio **BL:ND**

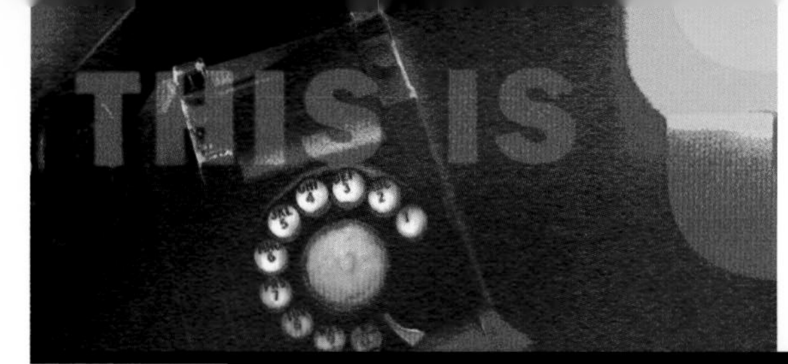

:30 PSA **Meth 911** | Creative Director/Editor/Animator **Chris Do** Designer/Animator **Jessie Huang**

Meth 911

Capturing the raw emotion of a 911 call, the spot uses edgy, graphically-charged typography to create a mood of acute agitation and distress. Fragments of gritty, abraded text are juxtaposed to make visceral the voiceover's frantic call for help, conveying an all-pervasive sense of desperation and helplessness. The dramatic contrasts of darkness and light which characterize the piece, were created by overlaying materials of varying grades of translucency (e.g., acetate, vellum, tracing paper, etc.) on a portable light box. The elements generated by the shoot were then cut together on an Avid Express, then finished in Adobe After Effects, where extra lighting and film grain were added.

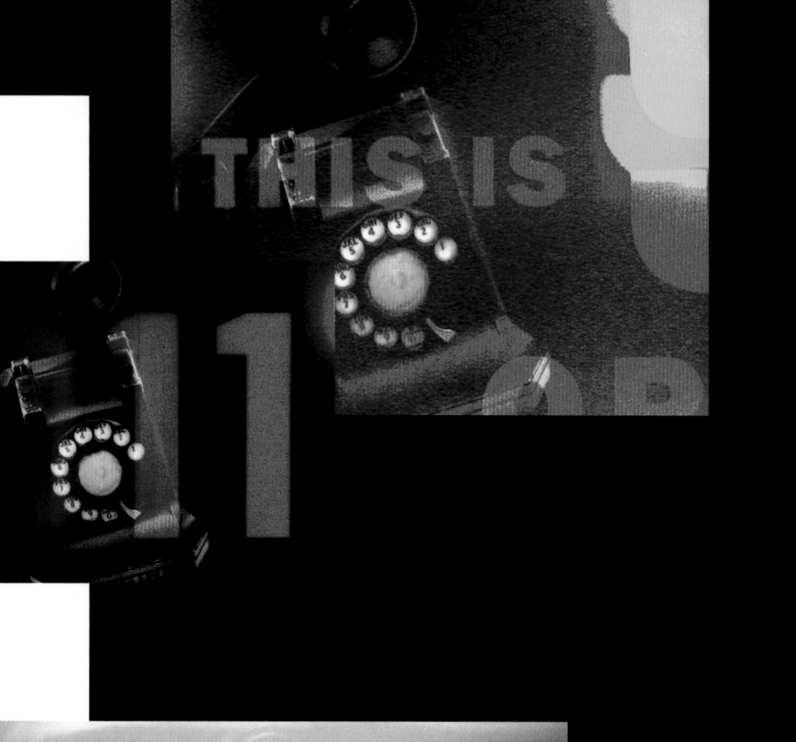

Photographer **Thai Khong** Studio **BL:ND**

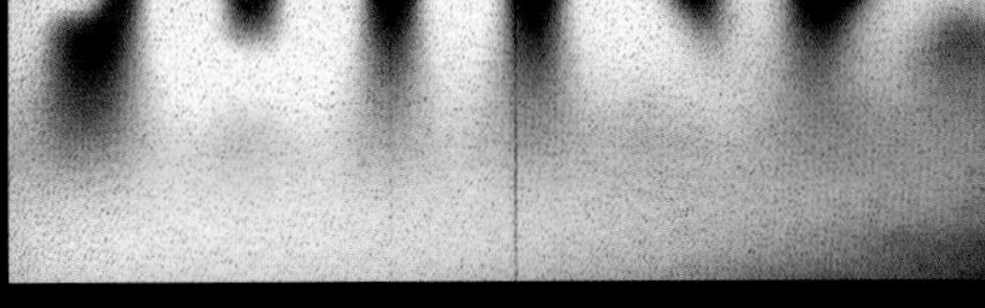

Main Title Sequence **Crank Calls** | Creative Director **Chris Do** 2-D / 3-D Animator **Lawrence Wyatt** 2-D Animator

Crank Calls

BL:ND created this main title with a dramatic macro view of a telephone dial sweeping through the frame. The titles are revealed through the finger holes of the phone's glowing dial. The shutter-like motion of the recoil generates a propulsive energy that drives the sequence forward and ends with the phone's ominous transformation into the cylinder of a loaded revolver. Using both digital stills and style frames for reference, the minute-long sequence was created entirely in 3d Studio Max, allowing greater precision and control over the complex, strobe-like animations. Following an offline edit, the final CG elements were taken to the Flame where they were further refined and then composited with the type animations.

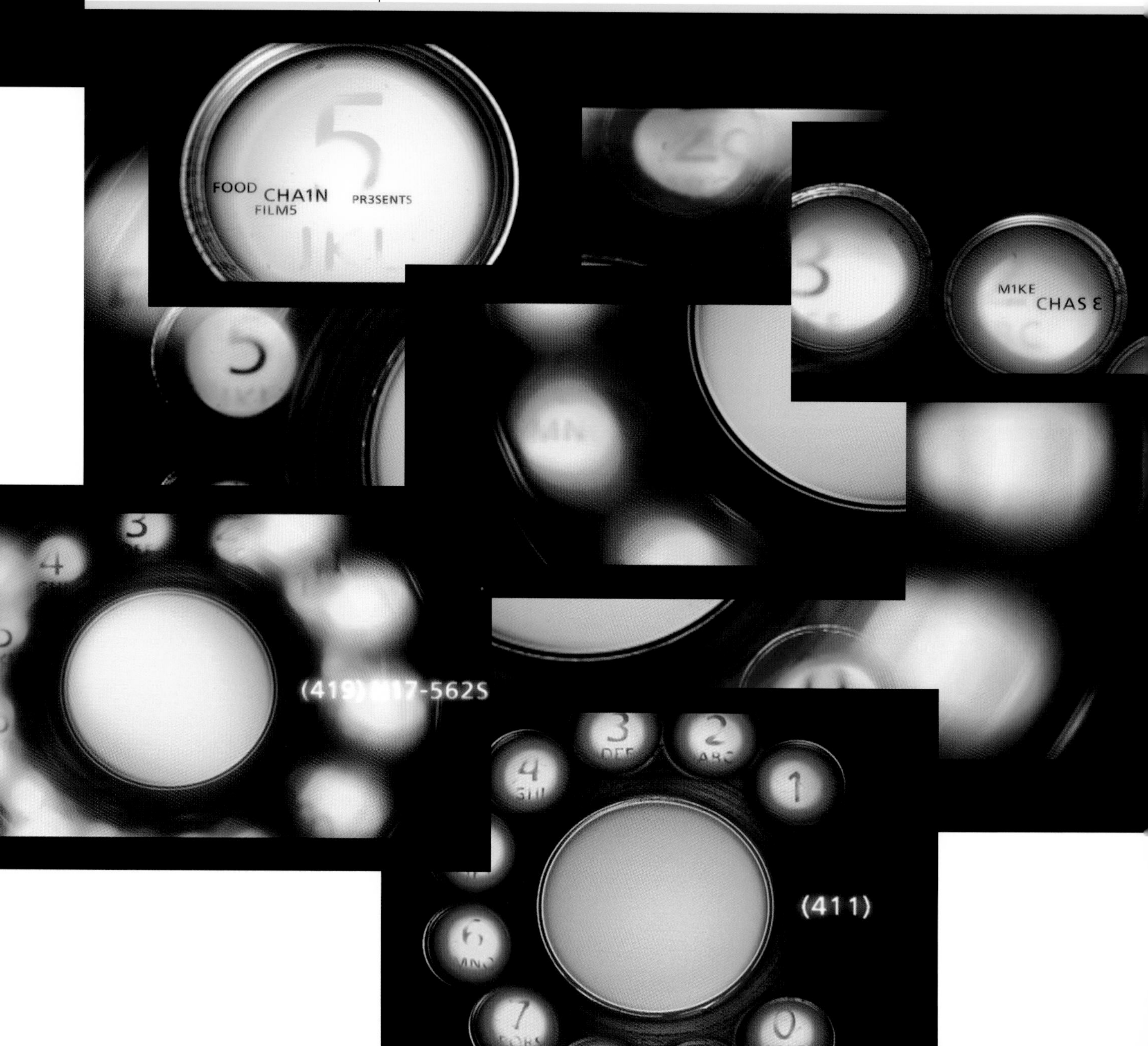

David Kerman Designer**Steve Pacheco** Photographer**Rick Spitznass** Editor**Erik Buth** Flame Artist**Lawrence Wyatt** Studio**BL:ND**

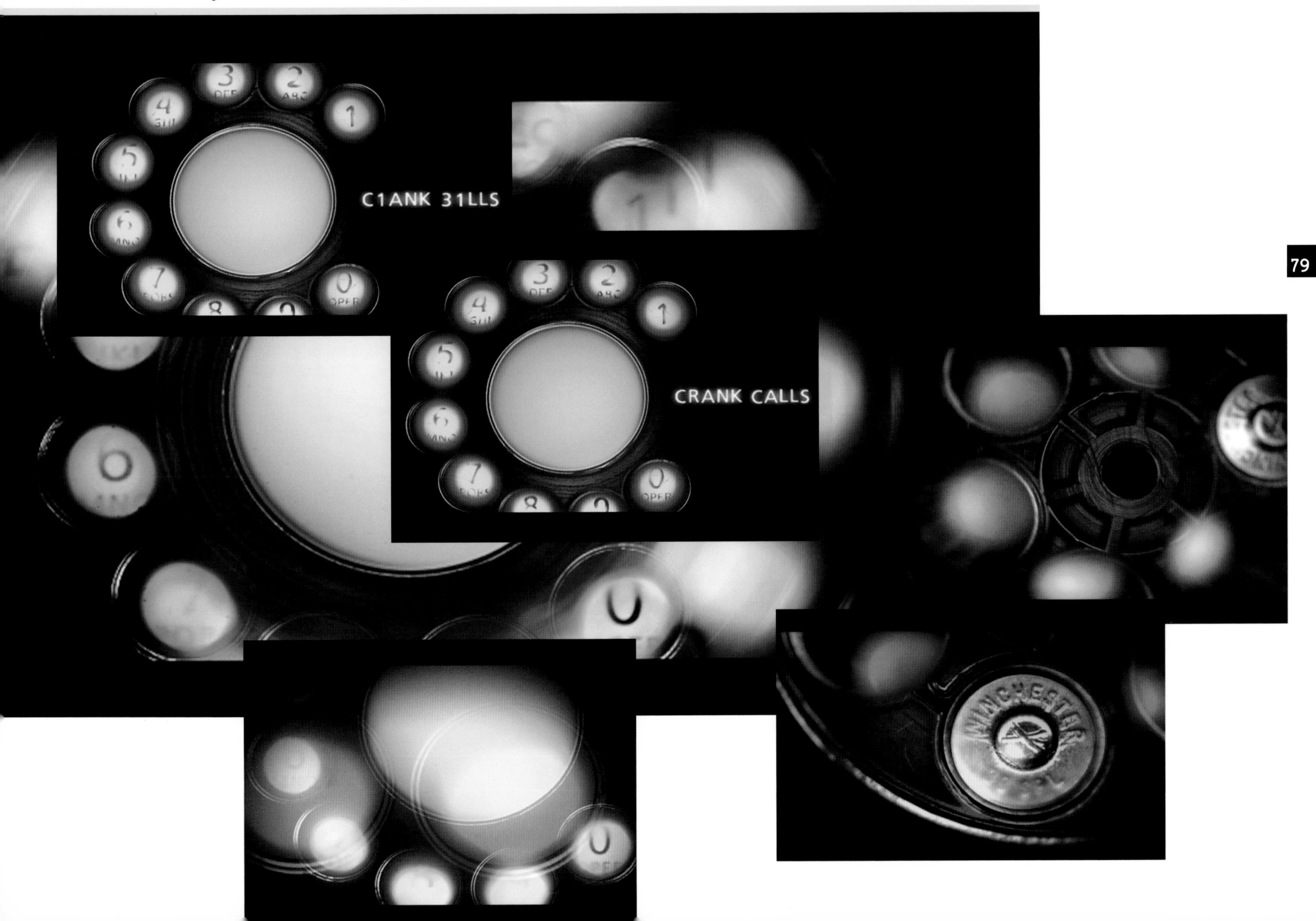

Network ID**CNN Ruby ID** | Creative Director**Garson Yu** Corporate Creative**Chris Moore**

CNN Ruby ID

CNN asked yU+co. to design a contemporary ID brand that speaks to a broader cross-section of viewers, targeting young adults. The creative team's solution was a three-dimensional, ruby-tinted glass rendering of the network's logo. Elegant, dynamic and light to the point of weightlessness, the execution suggests a lens through which to view the world. This treatment marks a sharp departure from the conservatism of previous CNN IDs. The richly colored, concept-driven piece was modeled and animated using Alias Wavefront Maya and composited with Adobe After Effects and Discreet Inferno.

3D Animator **Bryan Thombs** Studio **yU+co.**

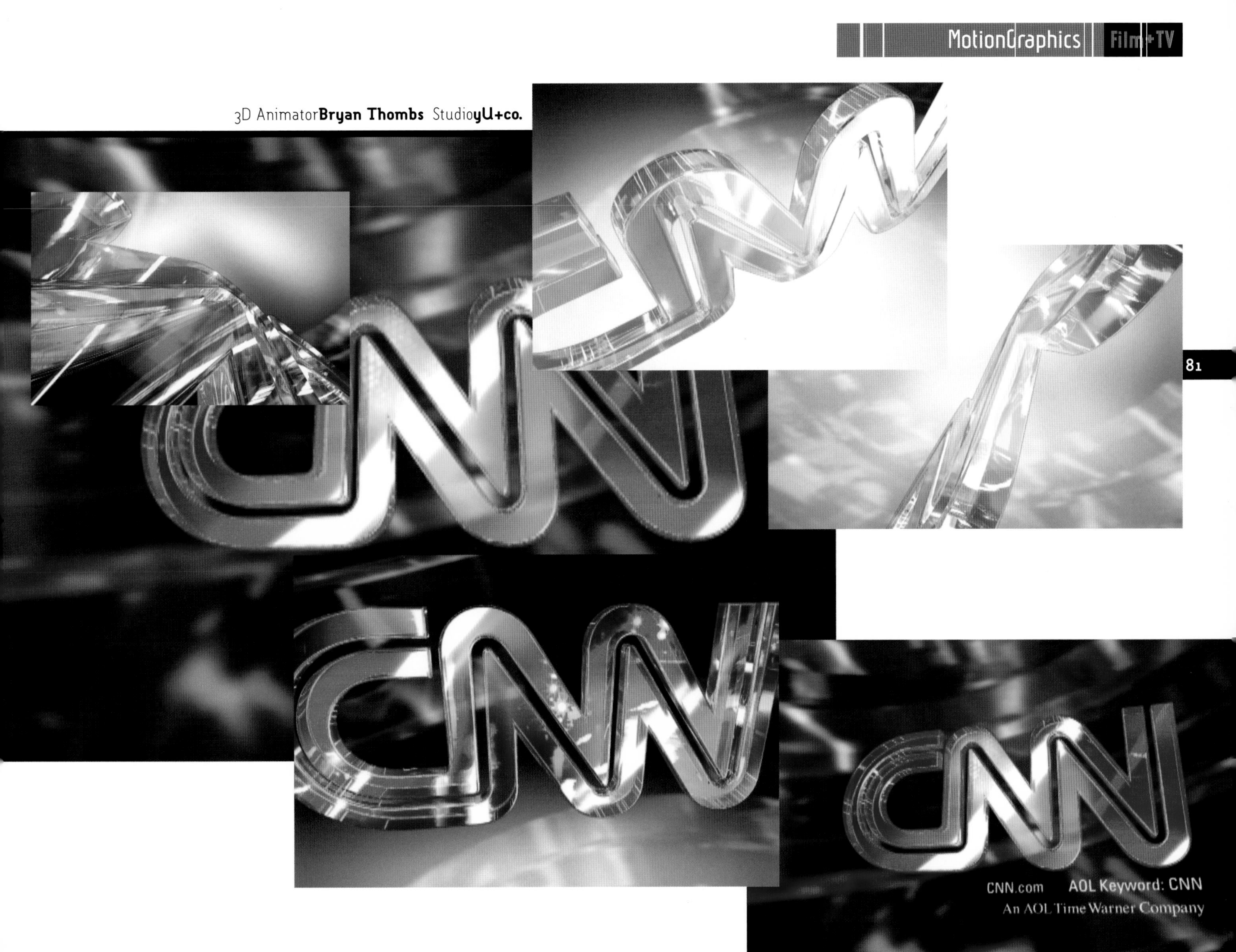

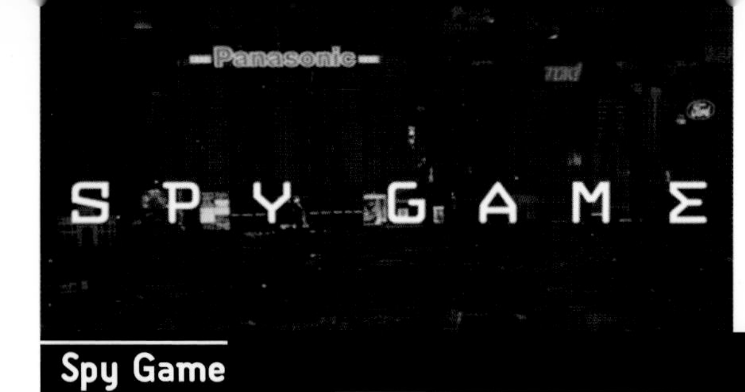

Main Film Title **Spy Game** | Designer/Creative Director **Garson Yu** | Designers/Animators **Benjamin Cuenod, Etsuko Uji**

Spy Game

yU+co. designed and produced the main title for the Tony Scott's stylish thriller, "Spy Game." Obscure lines of type, appearing over the film's first scene, decode to form the movie title and credits. The half-formed type suggests coded messages, while the furtive entry and exit of the typographic elements supports the film's escape theme. The creative toolset included Adobe After Effects and Discreet Inferno. The "Spy Games" opener exemplifies the yU+co. style of working closely with filmmakers in crafting main titles that are integral components of their work.

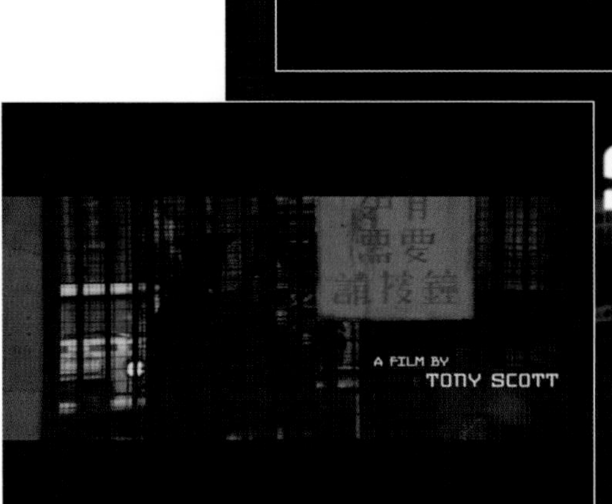

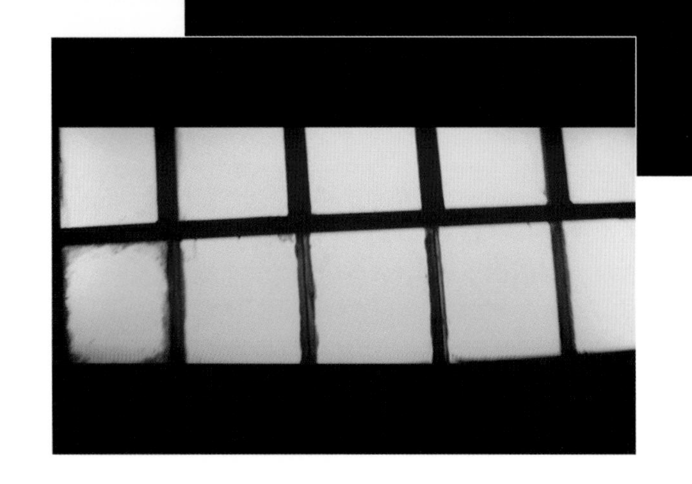

Typographer **Martin Surya** Inferno Artist **Todd Mesher** Studio **yU+co.**

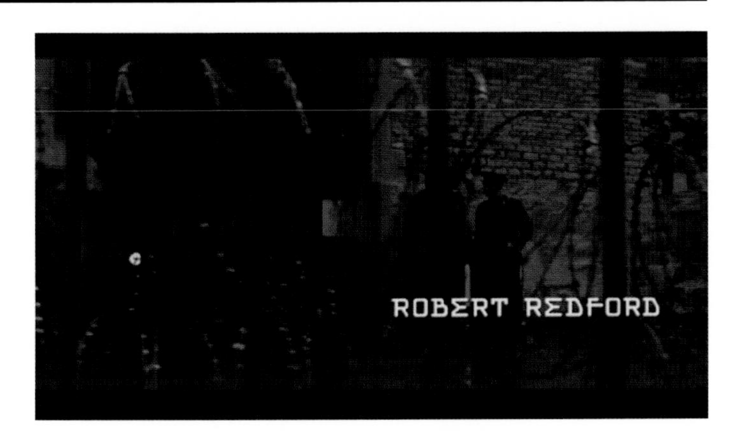

HBO Building Blocks

As part of HBO's overall network brand strategy, the HBO Building Blocks menu system serves to deliver a major portion of HBO's identity. Verb's goal was to create an organic, filmic and rich environment that would endure at least two years of repeated viewing. Streaming light and type were designed in Adobe After Effects as a 2D animated layer, then recomposited and animated in 3D space using AE for camera moves and depth of field. The liquid effect of the light streaks was created by using a liquid footage layer to distort a colorful abstract layer of footage. Background color splashes were made using solid colors in a similar technique.

HBO VP of Brand Image **Marc Rosenberg** HBO Creative Director **Karen Sands** Studio **Verb.**

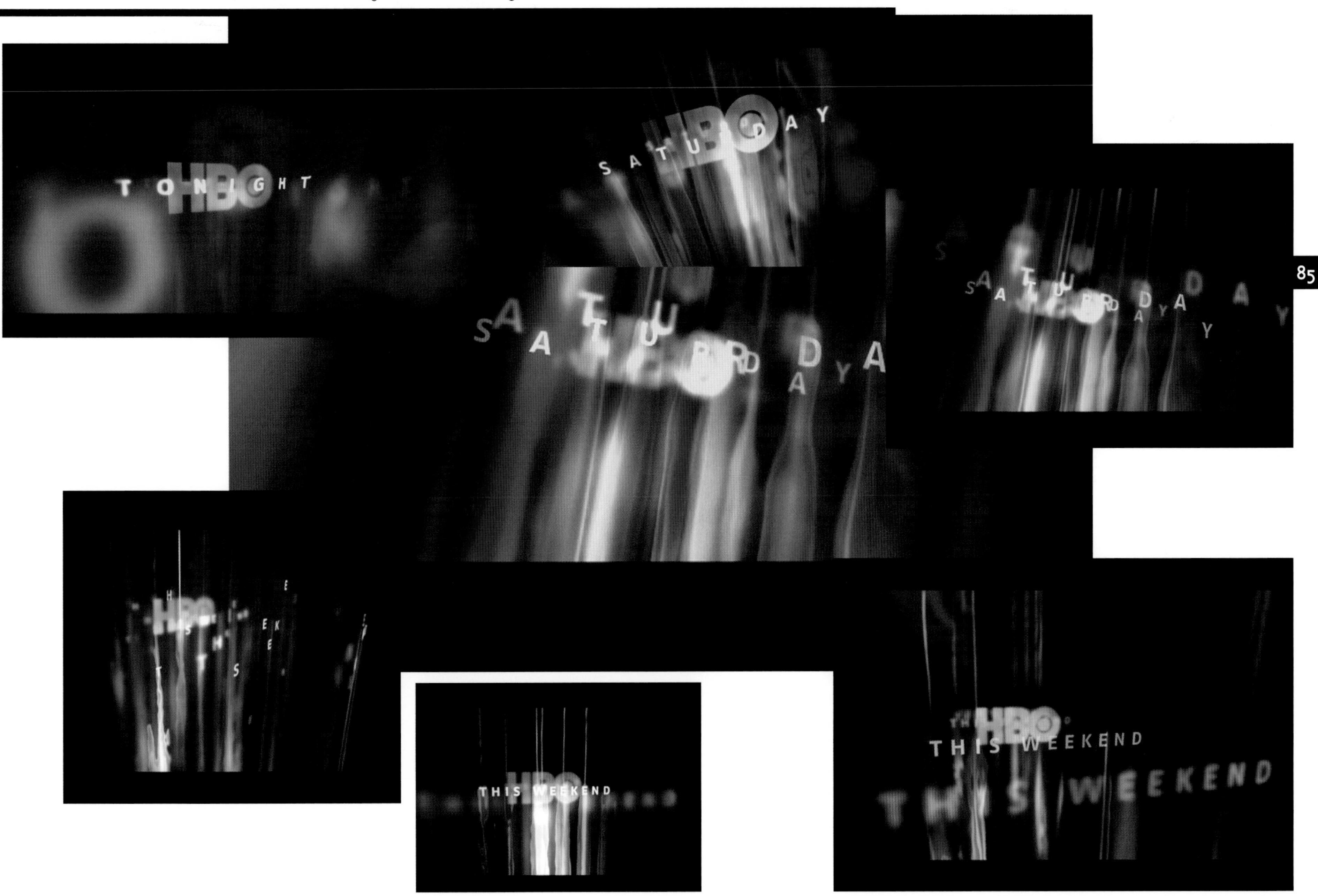

HBO Sex and the City "Cosmo"

The goal with the HBO Sex and the City "Cosmo" Tease was to build anticipation for the new summer season. Verb created a spot that was all about summer, fun and fashion: a brightly-colored area reveals themed icons like bikinis and bubbles, and concludes with the shape coming into focus as a cosmo-filled martini glass. The art was sketched by hand, then rebuilt for animation in Adobe Illustrator. The animation was built entirely in Adobe After Effects.

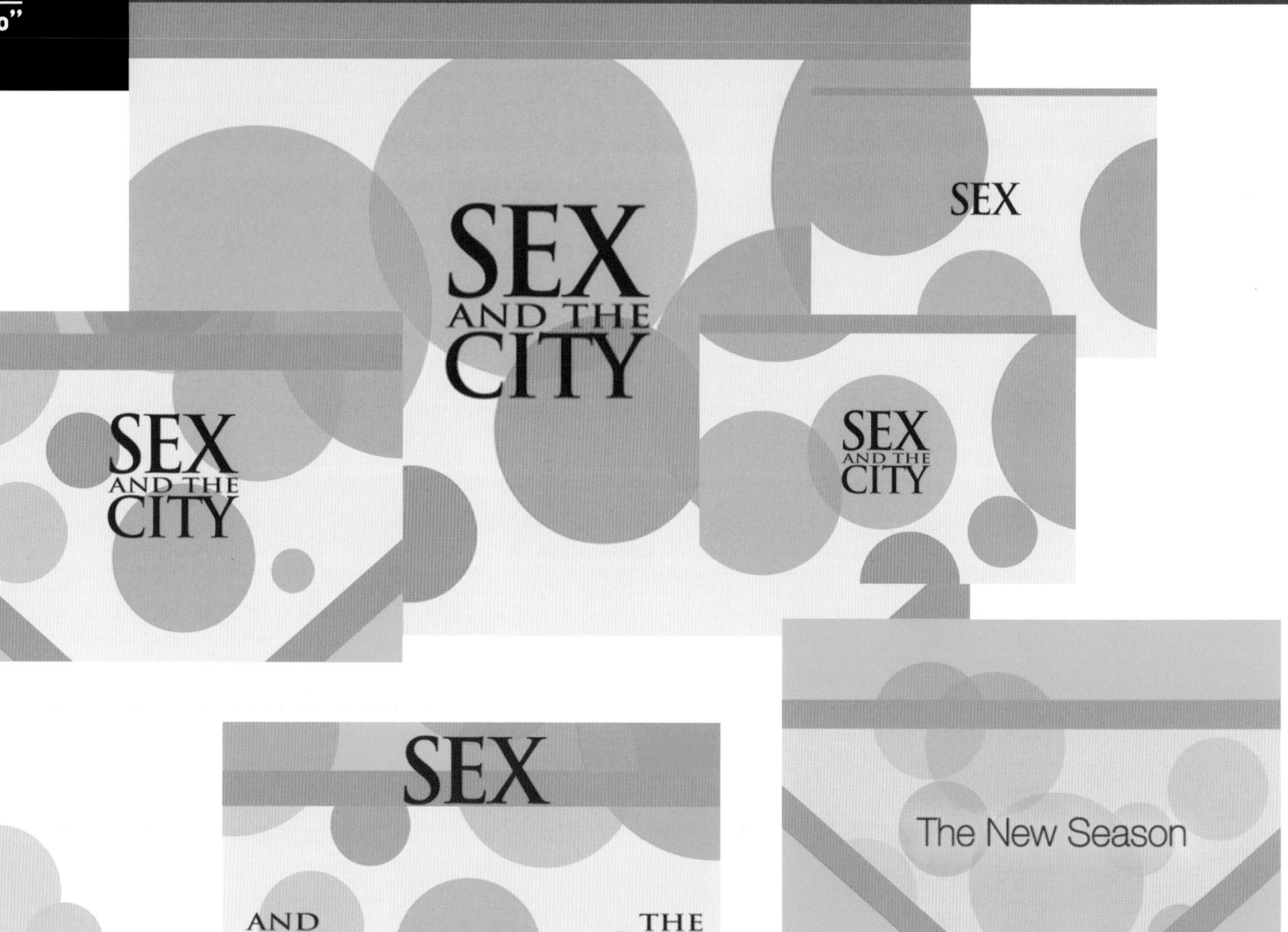

HBO Producer**Cami Errante** Studio**Verb.**

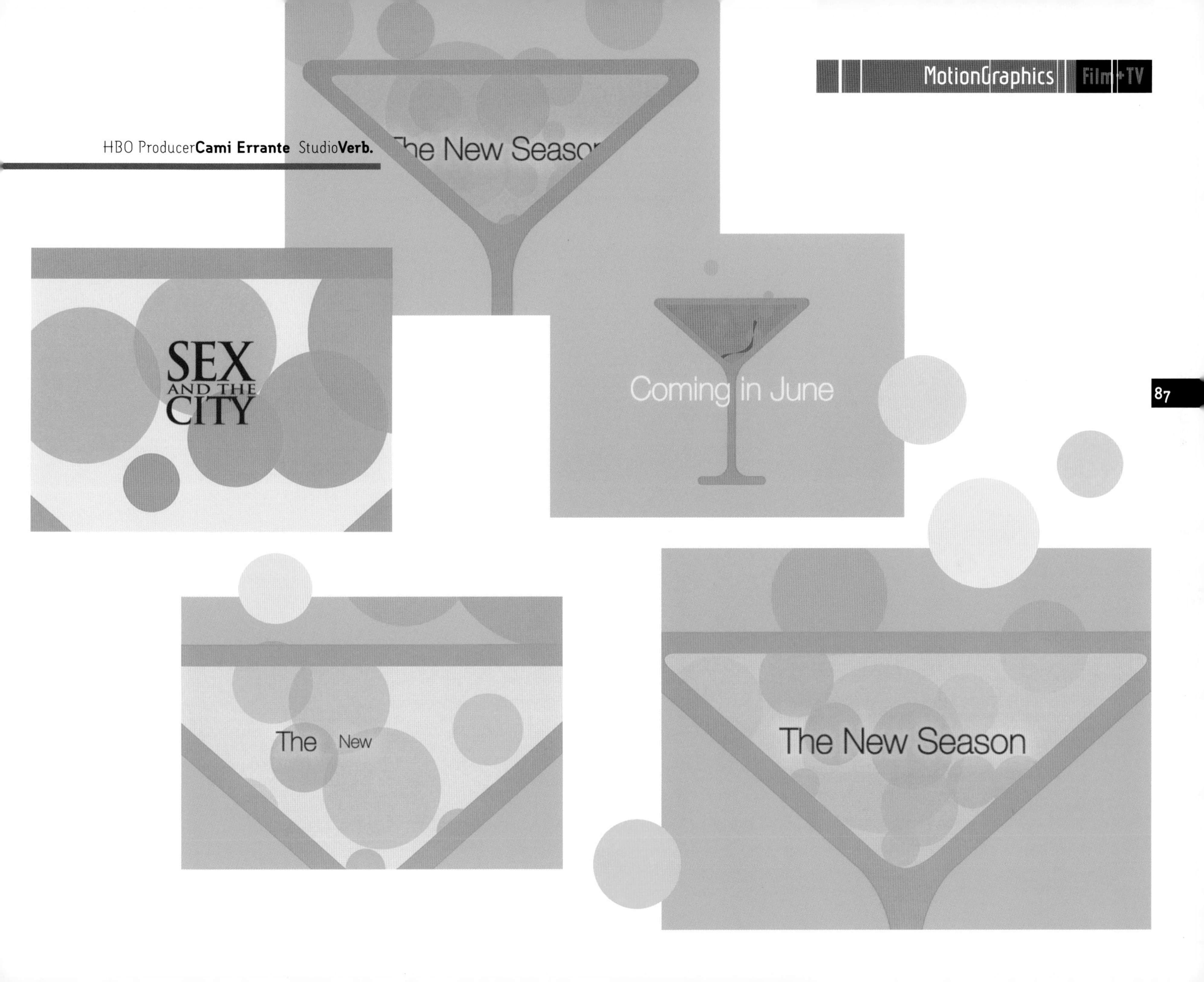

On-air Promo **HBO Critics 2000** | Creative Director **Greg Duncan** Producer **Bill Bergeron**

HBO Critics 2000

For HBO Critics 2000, the Verb team graphically played off the idea that the list of accolades for HBO content is limitless and non-stop: everywhere you look, critics are praising HBO. The solution is a series of vertically streaming "best-of" lists that create an abstract filmstrip composited with clips from the shows. One of the tricks in this spot was to use two identical layers of the graphic artwork; one layer of motion blurring in the animation and one not, resulting in a fast-moving blurred image, and the sensation of a quick-cut stutter effect.

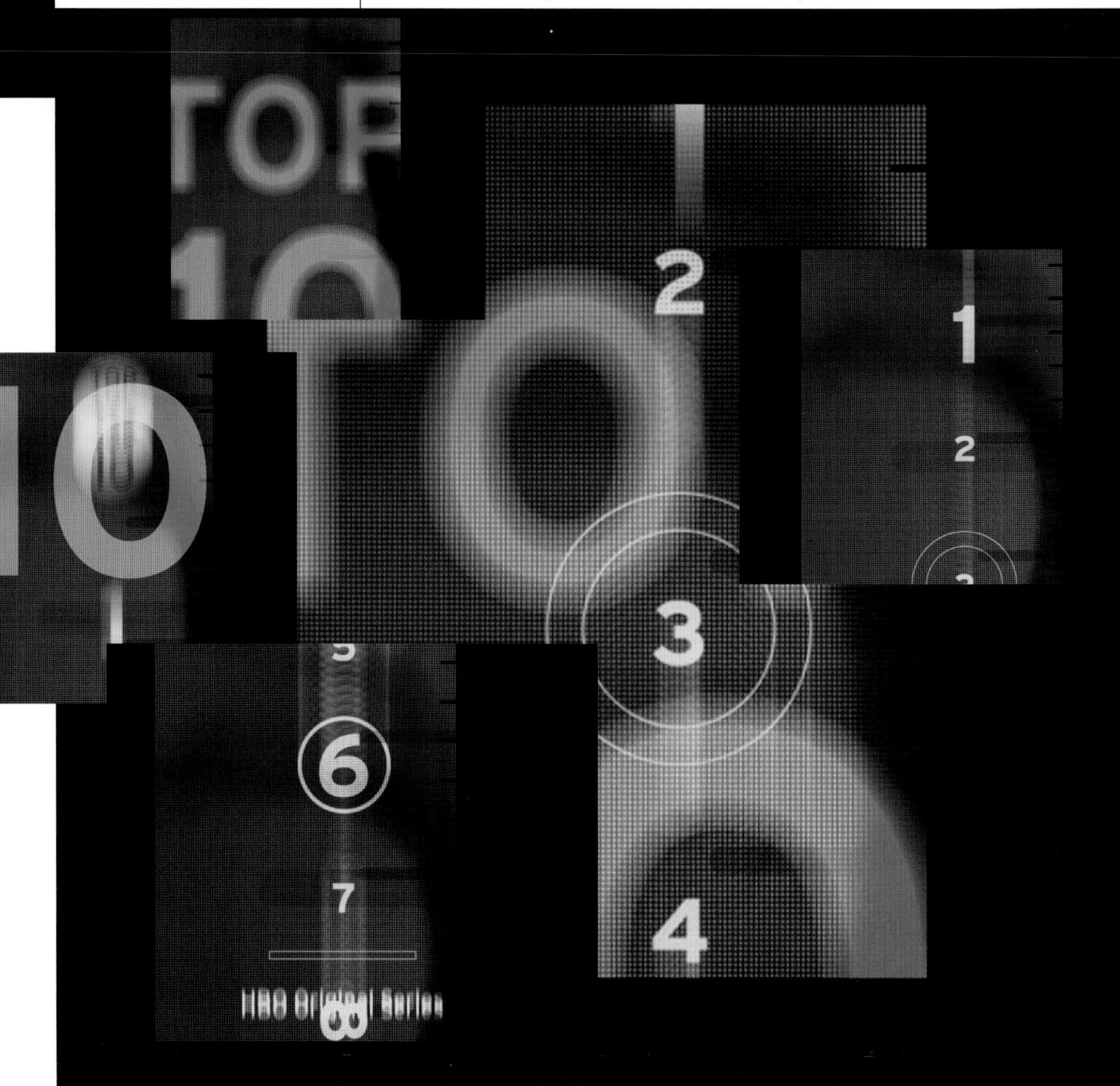

HBO VP of Brand Image **Marc Rosenberg** HBO Creative Director **Karen Sands** Studio **Verb.**

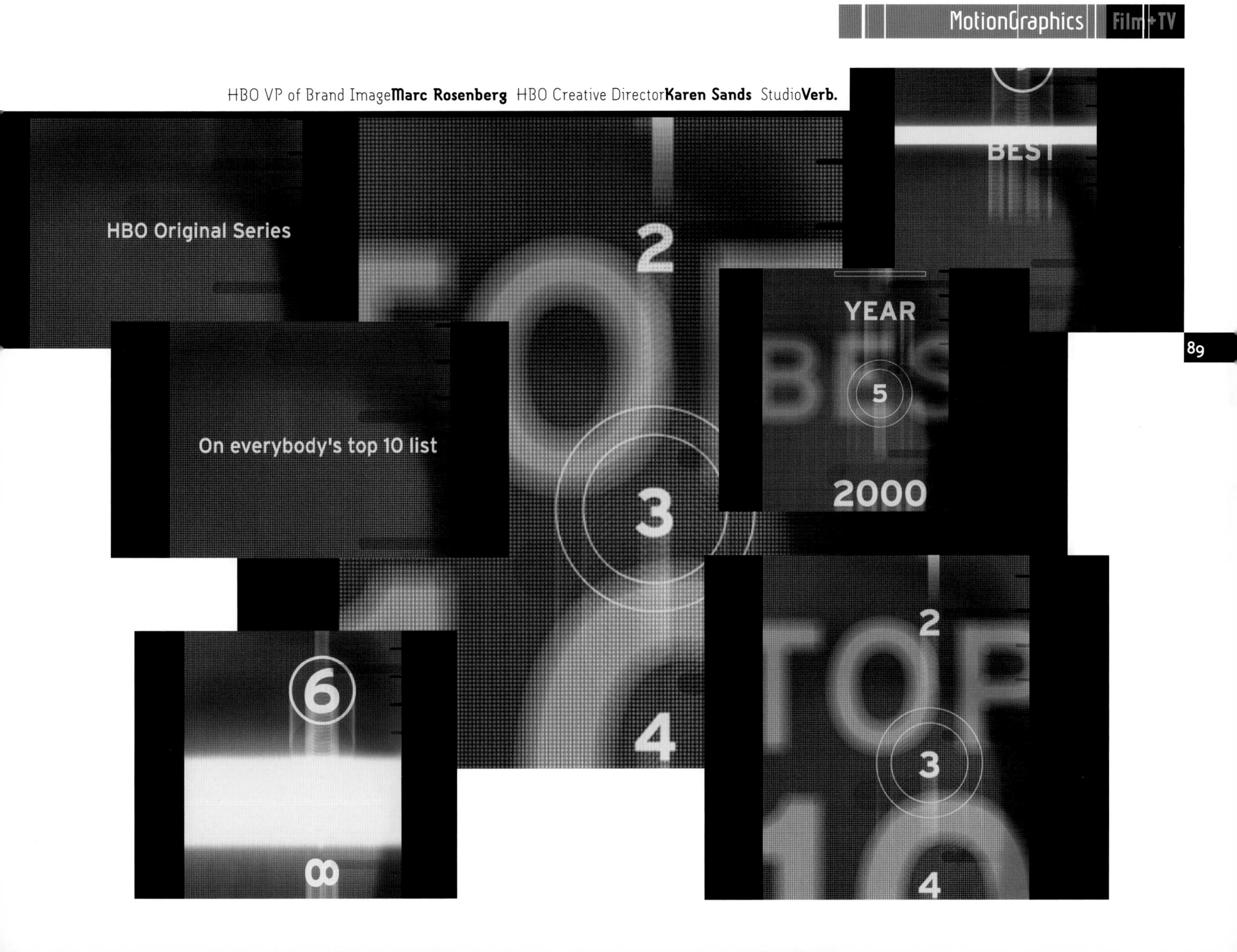

Franchise Package **HBO Sunday Night Series** | Creative Director **Greg Duncan** Producer **Bill Bergeron**

HBO Sunday Night Series

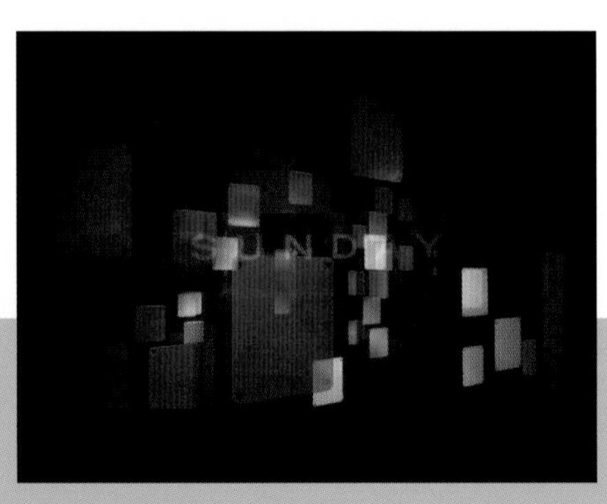

For HBO Sunday Night Series, the package needed to have the importance and gravity of film titles, but also required the cutting-edge immediacy of shows like the Sopranos or Oz. Verb designed a package centered around a series of animating graphic panels which transition into an abstract movie screen playing a montage of slow-motion film imagery. The animating panels and type were built in Cinema 4D. Color, depth and texture were delivered using movie footage, which was projected in 3D on the panels as they animate and lock into place to form a screen. Finish and effects were then added in AfterEffects.

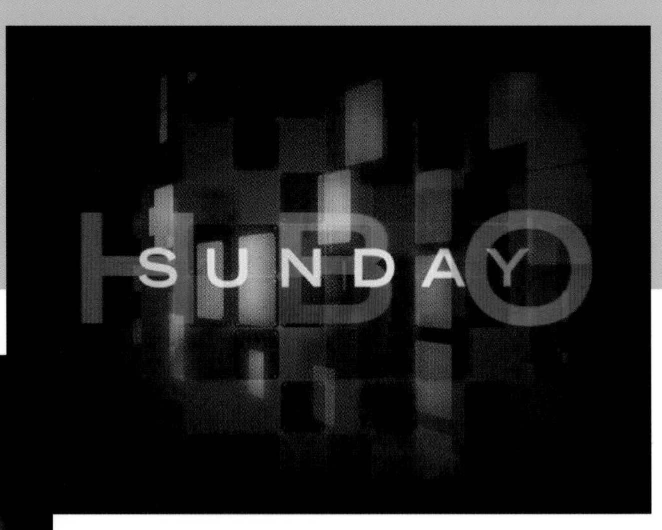

HBO VP of Brand Image **Marc Rosenberg** HBO Creative Director **Karen Sands** Studio **Verb.**

Cinemax Maxtime

The Cinemax Maxtime package had to say "film" without resorting to cliché. Verb's solution was to create a series of abstract graphic strips that animate to deliver words and logos describing Maxtime. The strips were designed to feel vaguely evocative of filmstrips, but also to appear like a liquid surface where light and graphics interact organically. The spots were created by building the filmstrip animations in Adobe After Effects as high-resolution QuickTime files, which were then surface-mapped onto 3D filmstrips and animated in Cinema 4D, and subsequently recomposited in Adobe After Effects.

HBO VP of Brand Image **Marc Rosenberg** Cinemax Creative Director **Robert Priday** Studio **Verb.**

SAP Formula 1 Campaign

Verb's challenge in creating the SAP Formula 1 Campaign was to tell an extended narrative using type onscreen without voice over support. The spot also had to stand out against many other Formula 1 promos that would be utilizing similar footage. The solution was to create a rich graphic world in which the race car footage could live. The graphics keep up the pace, while allowing the typographic story to play out without feeling heavy-handed or difficult to follow.

Creative Director**Roger Woo/Wooart** SAP Producer**Mario Giampaglia** Studio**Verb.**

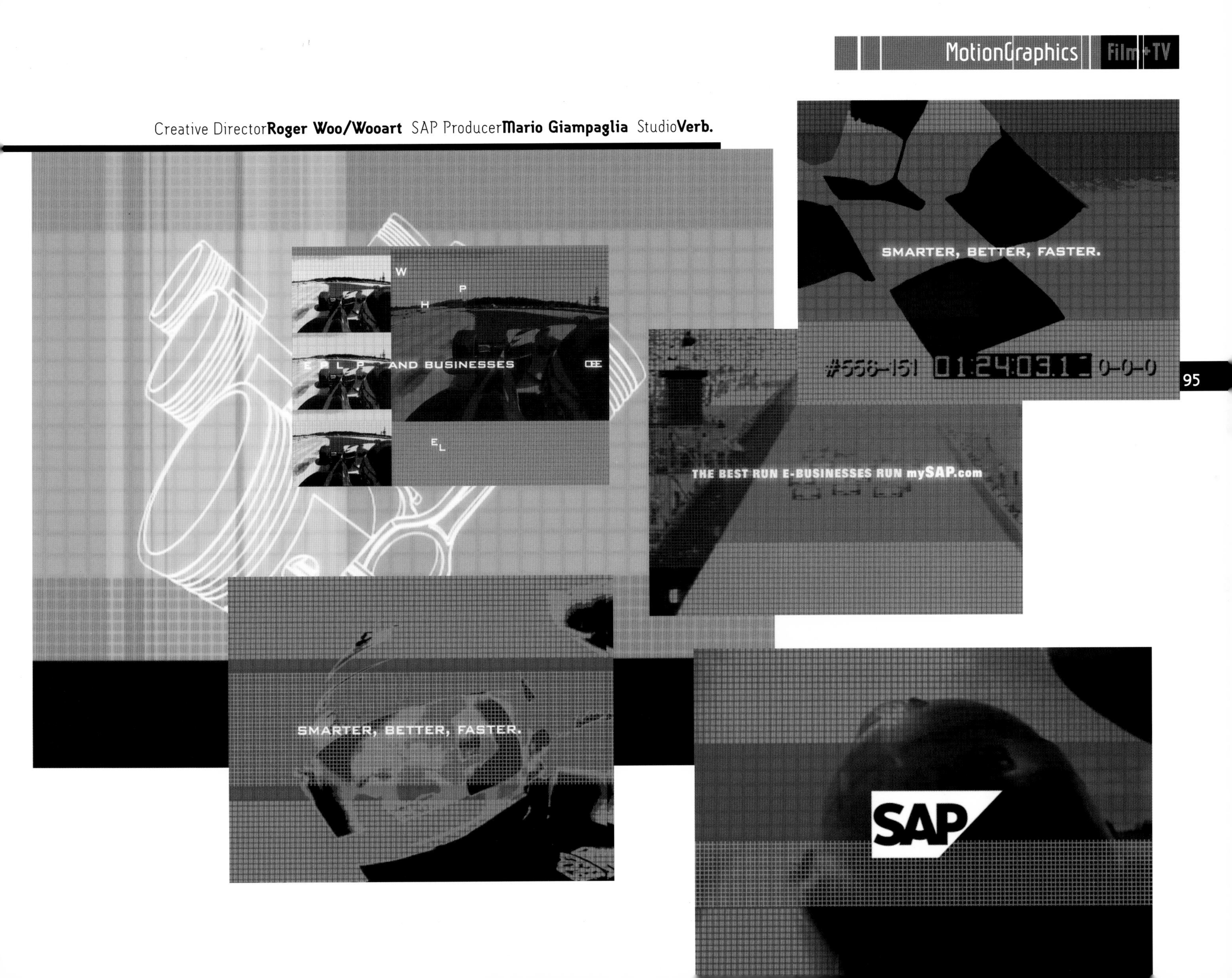

Verb Show Reel

Self-Promotion**Verb Show Reel** | Creative Director**Greg Duncan** Producer**Bill Bergeron** Studio**Verb.**

For their show reel, Verb wanted to extend the concept of "action" using the verb iconography that runs throughout the studio's identity. The Verb team turned the reel into an art project. Starting with their identity elements—logo, verb icons, shapes and color palette—they created a huge, high-resolution language consisting entirely of verbs. Using these hi-res images as a starting point, a series of animations were built frame by frame in Adobe Illustrator. The frames were then imported into AE to be sequenced, retimed and composited. What evolved was a series of iconic, typographic and graphic animations that serve as intro, interstitial transitions and conclusion to the reel.

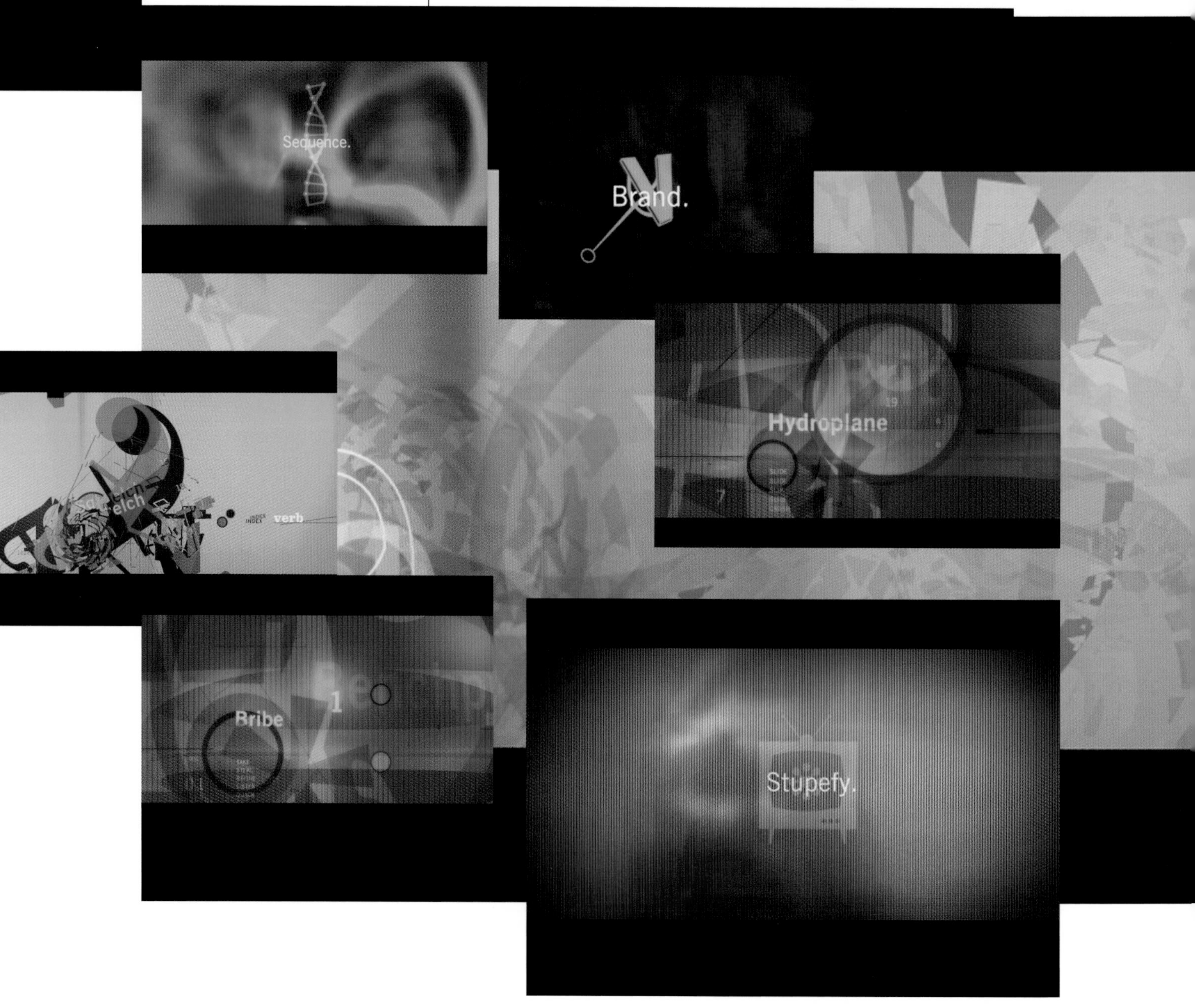

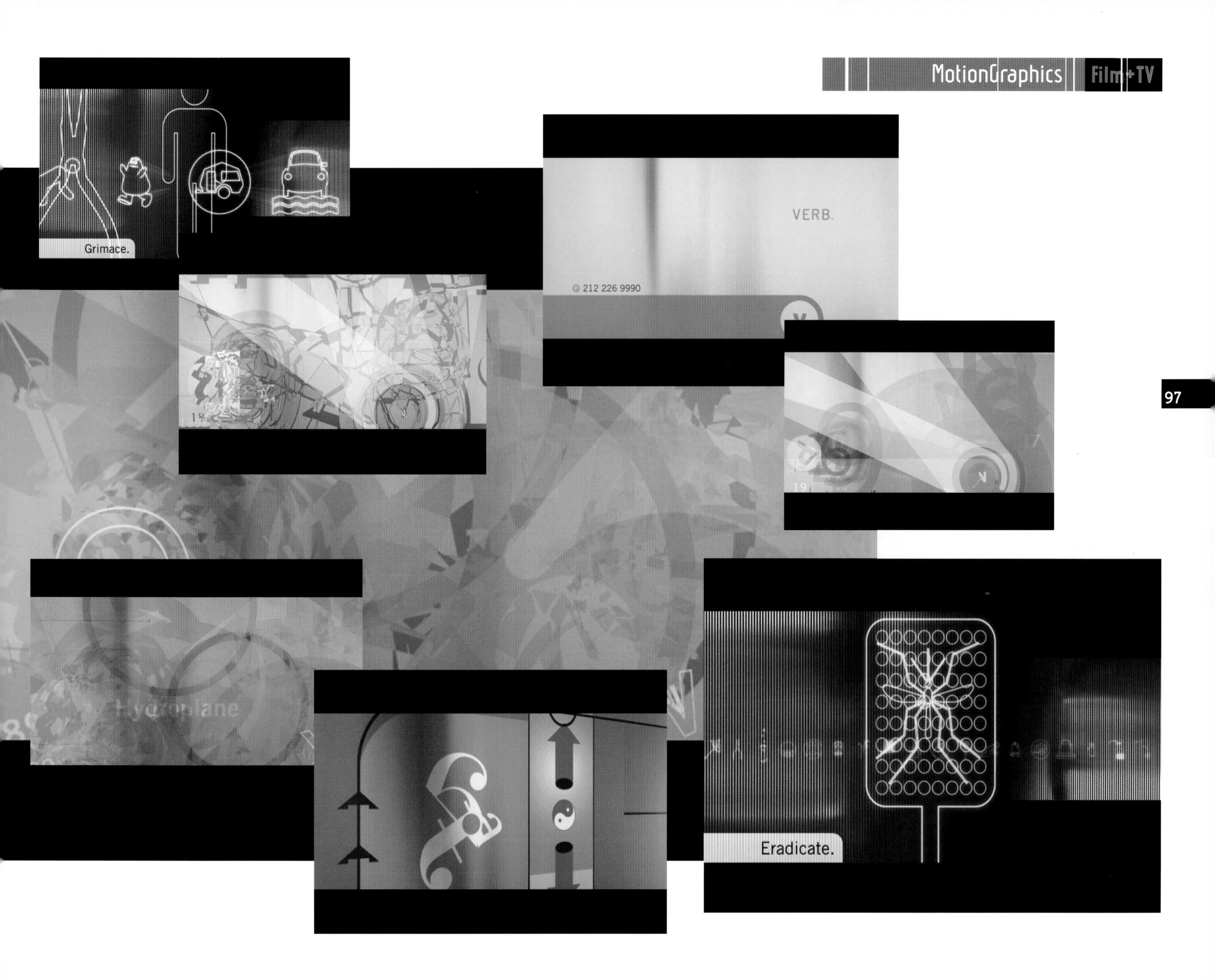
Grimace.
VERB.
212 226 9990
Hydroplane
Eradicate.

Show Open **Sundance Channel Shorts Stop** | Designer/Director **Julian Grey** Studio **Head Gear Animation**

Sundance Channel Shorts Stop

Head Gear's objective for Sundance's new opening for a show about short films, highlights the unique nature of the genre. Footage was created with Super 8, 16mm still cameras and Digital Video. Compositional and textural elements were shot frame-by-frame on a tabletop using backlit treated acetate. Adobe After Effects was used for final compositing and tinting. The spot opens on black to the sound of dripping water, revealing pools of expanding imagery, culminating in a blink preceding the titles. Circles were the primary motif, framing and transitional device. The shapes represent notions of time and diversity of approach, and more literally the eyepieces and lenses of film optics.

sundance
channel
SHORTS
STOP
STOP

Wamdue Video

Graphic Havoc developed this video for the world tour of Strictly Rhythm Records recording artist Wamdue Project. Designed for live musical performances, the piece uses hand-drawn elements, manipulated typography, original typefaces, still and animated 35mm photography and new video footage shot by Graphic Havoc. The studio also created the promotional material for the group's latest album, "Program Yourself." GHAVA developed the idea for the project based on motion and transit in an urban environment. Due to budget constraints, the designers needed to produce new images. They used 35mm animated stills which became the starting point for the entire project.

Studio**Graphic Havoc avisualagency**

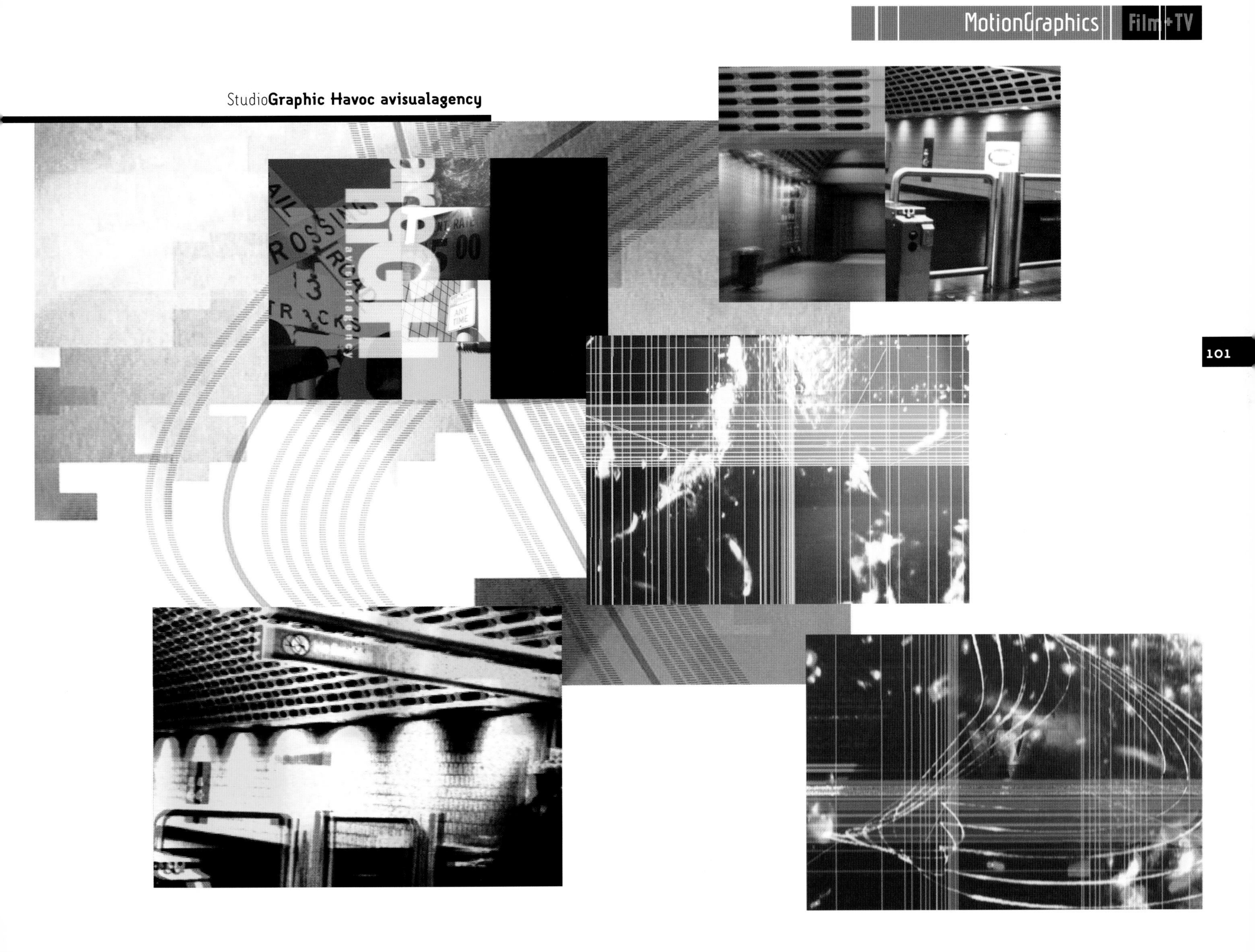

"Tema Composición: La Vaca"

"Tema Composición: La Vaca" is a project conceived by Argentina's Grupo Doma. The studio was searching for a design theme for a video and at the time, they were only painting cows in the streets of Buenos Aires. Someone jokingly suggested making a film entitled, "The Cow," akin to the simple compositions of early grade school writing exercises. The funny concept struck a chord, and the Doma crew worked non-stop over a weekend, creating in crazy-cow mode. The result is a comical study and exploration of a docile animal, and a continuation of Doma's non-commercial experimental works using the world as its laboratory.

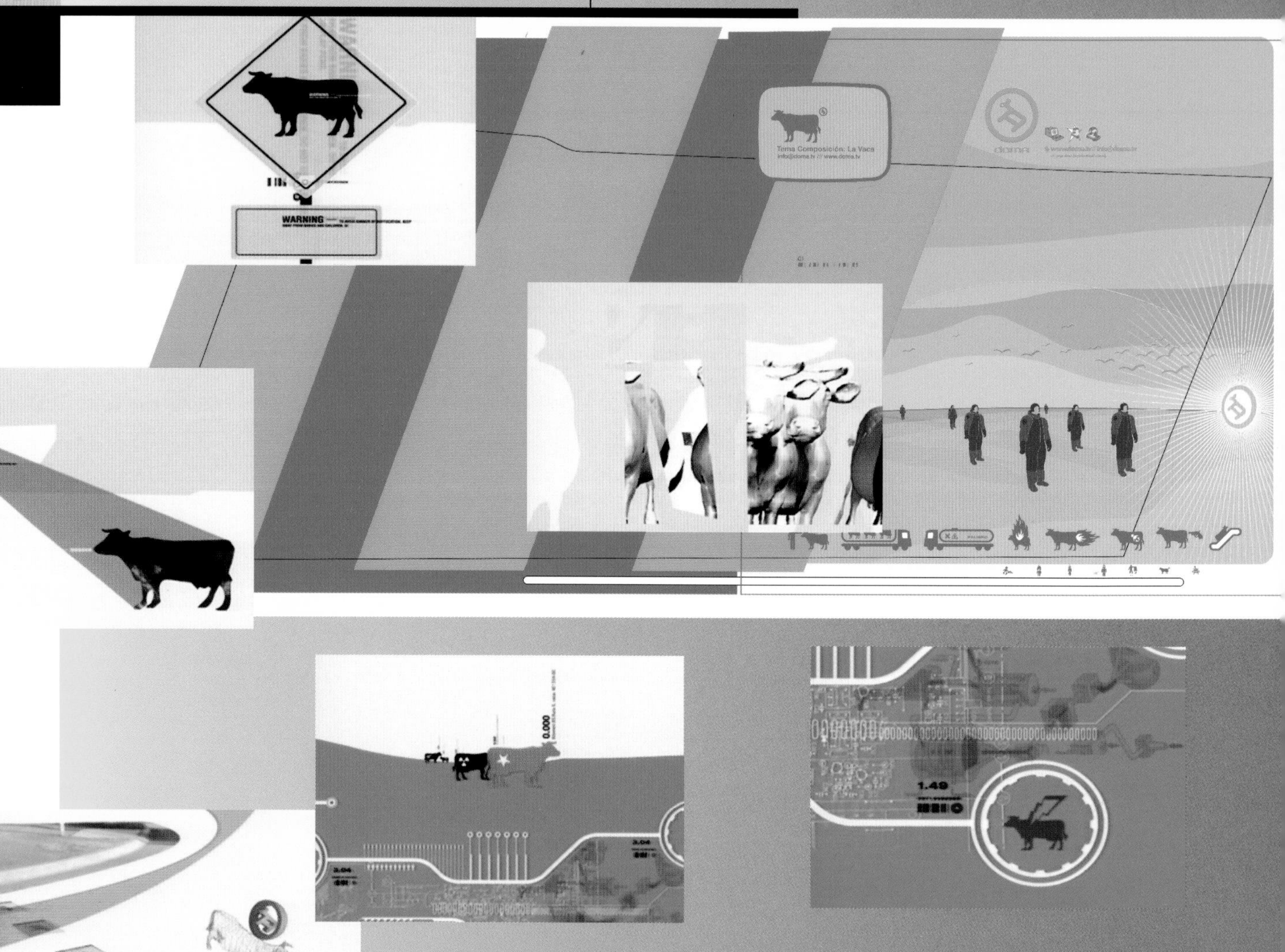

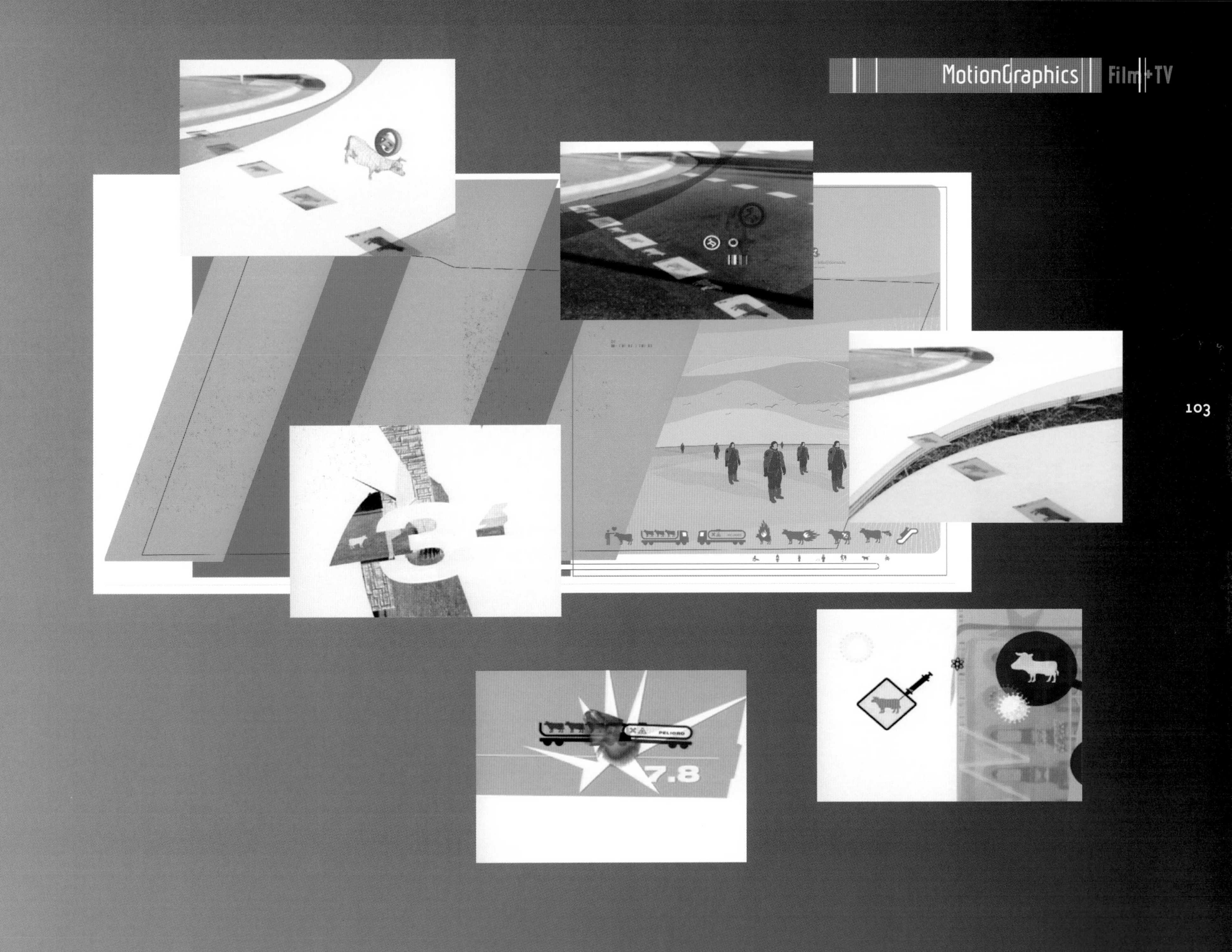
PELIGRO
7.8

45mph is a music video commissioned by the No Bones label to promote recording artists Genik. D-Fuse used the "road movie" as a central motif in the piece. The video examines beauty in the mundane, taking road markings and textures from nature and expanding them into new forms. The cutting and contrasts of color and form complement the mood of the music. The video was shown continuously throughout Europe and North America on many popular television networks including MTV and Viva. Additional screenings were held at festivals and events in Barcelona, Helsinki and other major European cities.

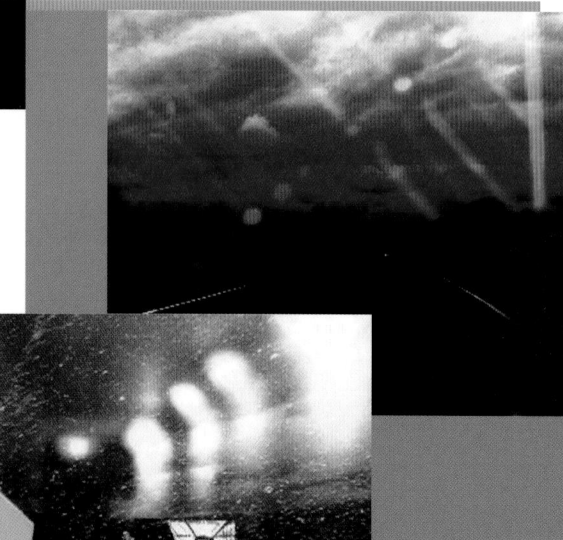

VideoEndeka | DesignerD-Fuse StudioD-Fuse

Endeka

The No Bones label called upon D-Fuse to create a video for Fluid. Endeka conjures a world in which abstract forms collide and re-define themselves as patterns of intricate complexity—their shape and texture reflecting the jazzy mood of the music. The music and imagery evolved simultaneously as part of the ethos of the D-Fuse DVD. A collection of rich and multi-textured graphic imagery was animated in Adobe After Effects and set to the music created by Dominic Glynn. Endeka was exhibited at the Avanto Film Festival in Helsinki and other major events throughout Europe. Shown on MTV and Viva, the video was also featured on DVDs produced by Gasbook, Creative Review and 72 dpi.

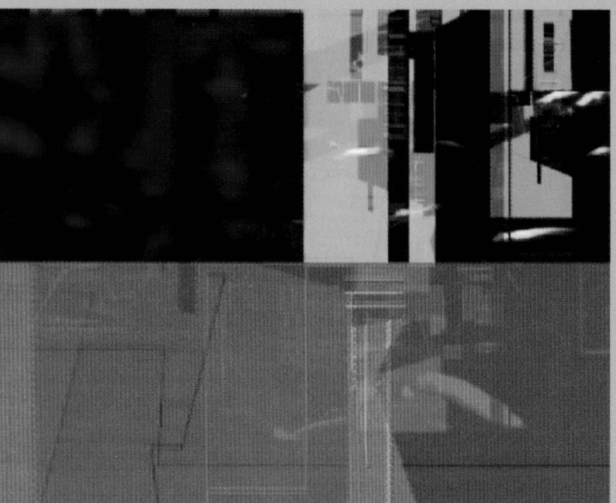

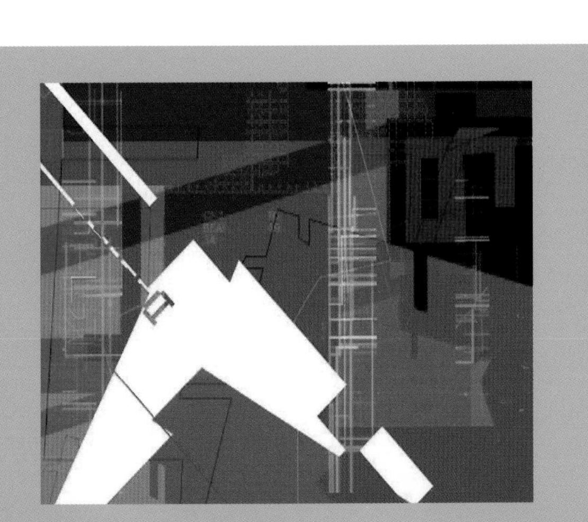

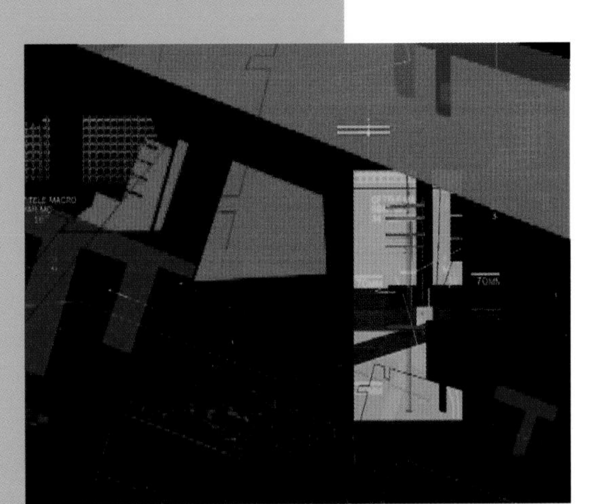

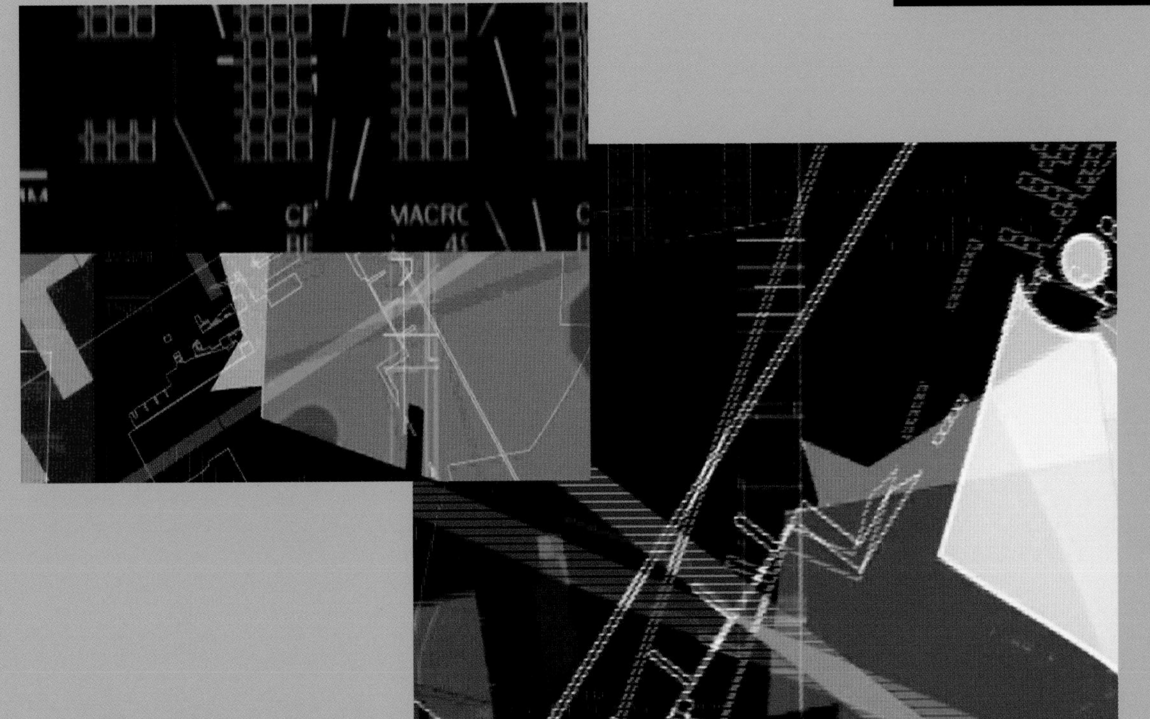
MACRO

XFM

Film Ident **XFM** | Designer **D-Fuse** Studio **D-Fuse**

D-Fuse created this identity piece commissioned by the UK national independent station XFM. A series of logos were treated with the use of hot and cold textures—fire and ice—to produce contrast and dramatic effect. The rich, dark textures of the backgrounds complement blurred, flaring and fading logo treatments. The work was transferred to video and 16mm film for use on tour events and in many promotions. The spot reinforces the edgy appeal of London's cool indie radio.

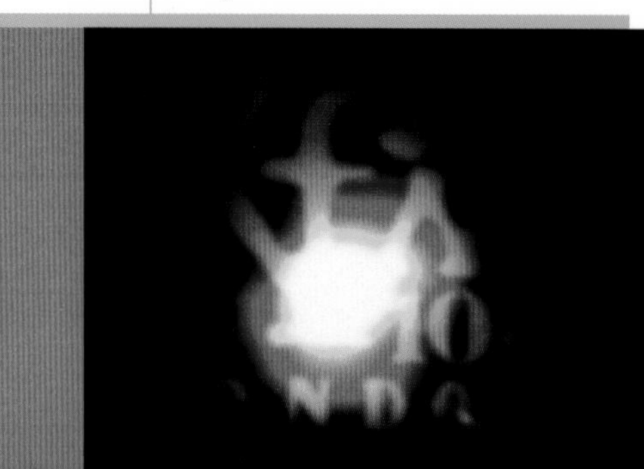

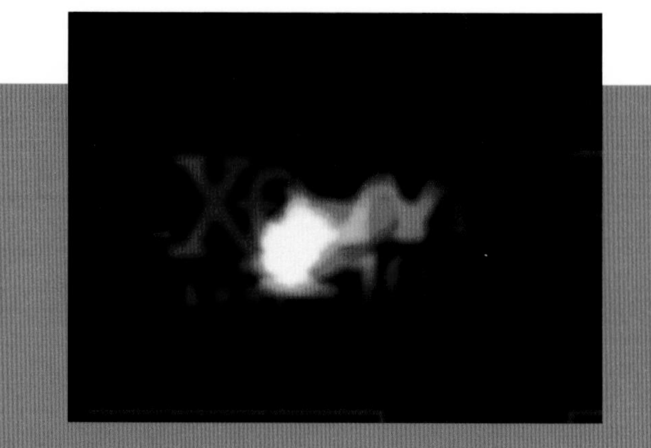

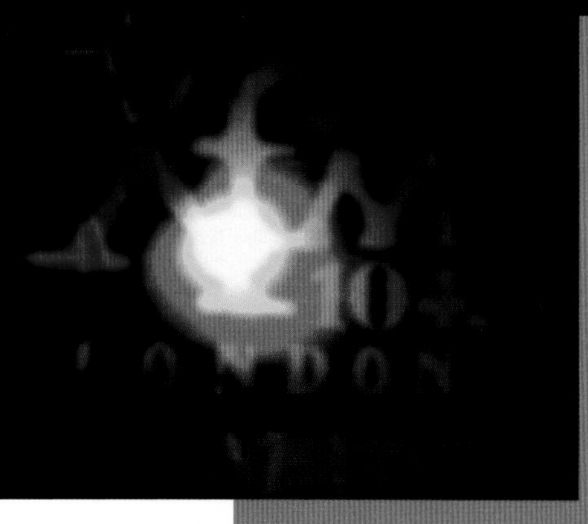

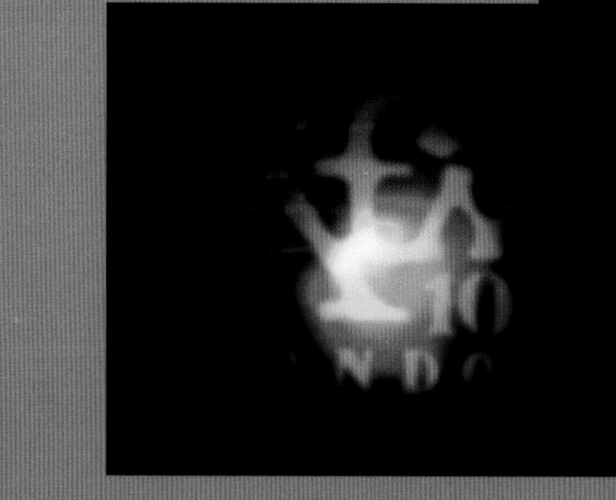

104.9
LONDON

DyDo Real Black

TV Commercial **DyDo Real Black** | Designer **Sei Hishikawa** Studio **DRAWING AND MANUAL**

DyDo asked Tokyo's Drawing and Manual to create a spot for their new line-up of Real Black coffee beverages. Drawing and Manual combined two-dimensional graphic design with 3D expression in developing the project. To make the "black" conspicuous, the designers utilized a series of contrasting patterns and a varied color palette. Morphing patterns and shifting colors yield to stark white animated typography over a black background. The piece attracts the audience with its "fashion color" theme, progressing from the trendy opening hues to the classic appeal of basic black.

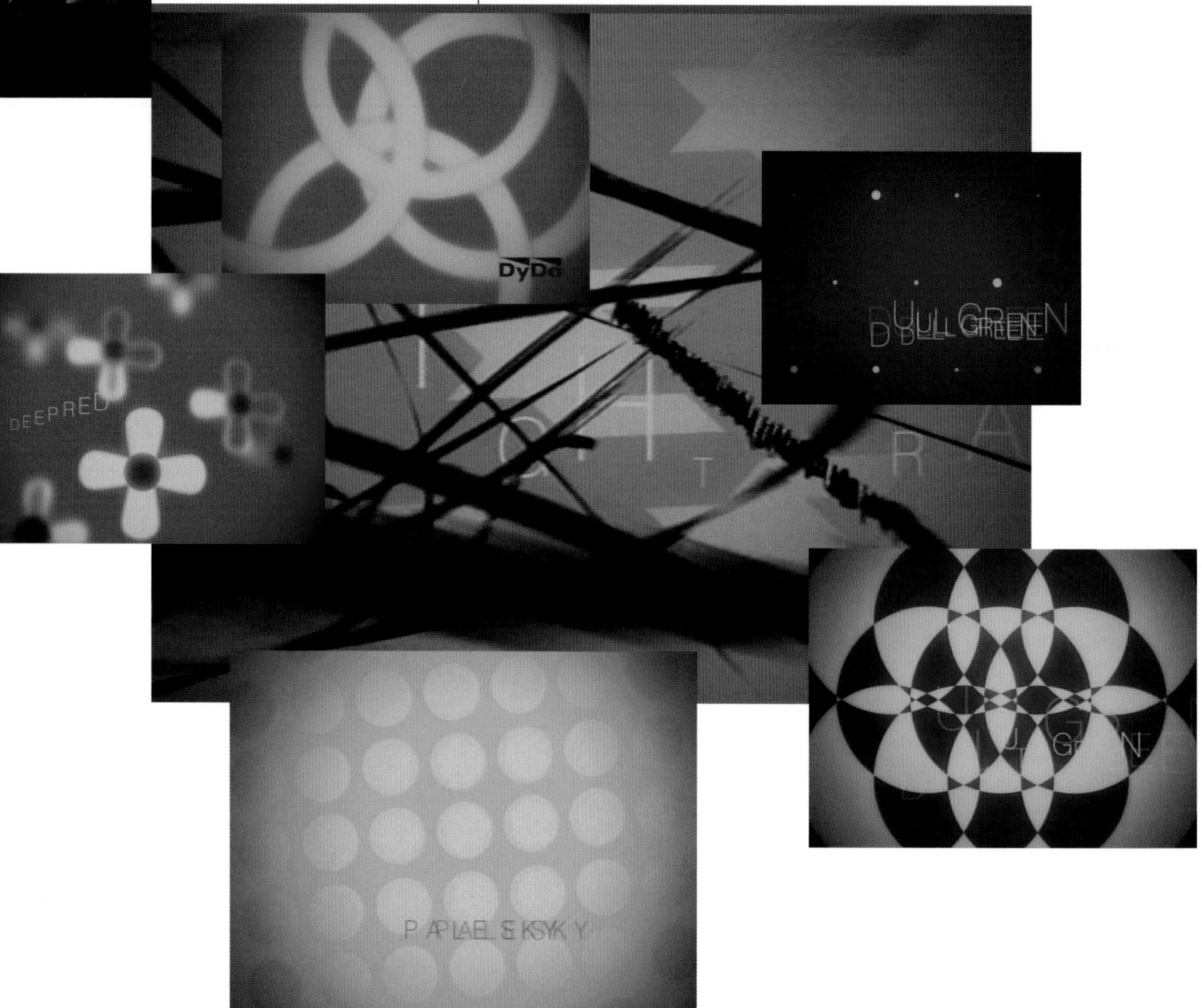

DyDo
BLEND
Coffee
リアルブラック
NEW LINE UP

HOTWIRED

Promotion Image **HOTWIRED** | Designer **Sei Hishikawa** Studio **DRAWING AND MANUAL**

Drawing and Manual developed this promotional piece for Hotwired, the online version of "Wired" magazine. Digital and experimental expression were key in the design of the project, and a primary goal was to break with the more conventional methods of showing images. Layer upon layer of imagery was built, with sequences moving rapidly through diverse elements relating to the popular internet publication. In some places, the piece is comprised of more than 600 layers. Surprisingly, once the initial concept was established, work on the energetic, complex project was completed in about eight hours.

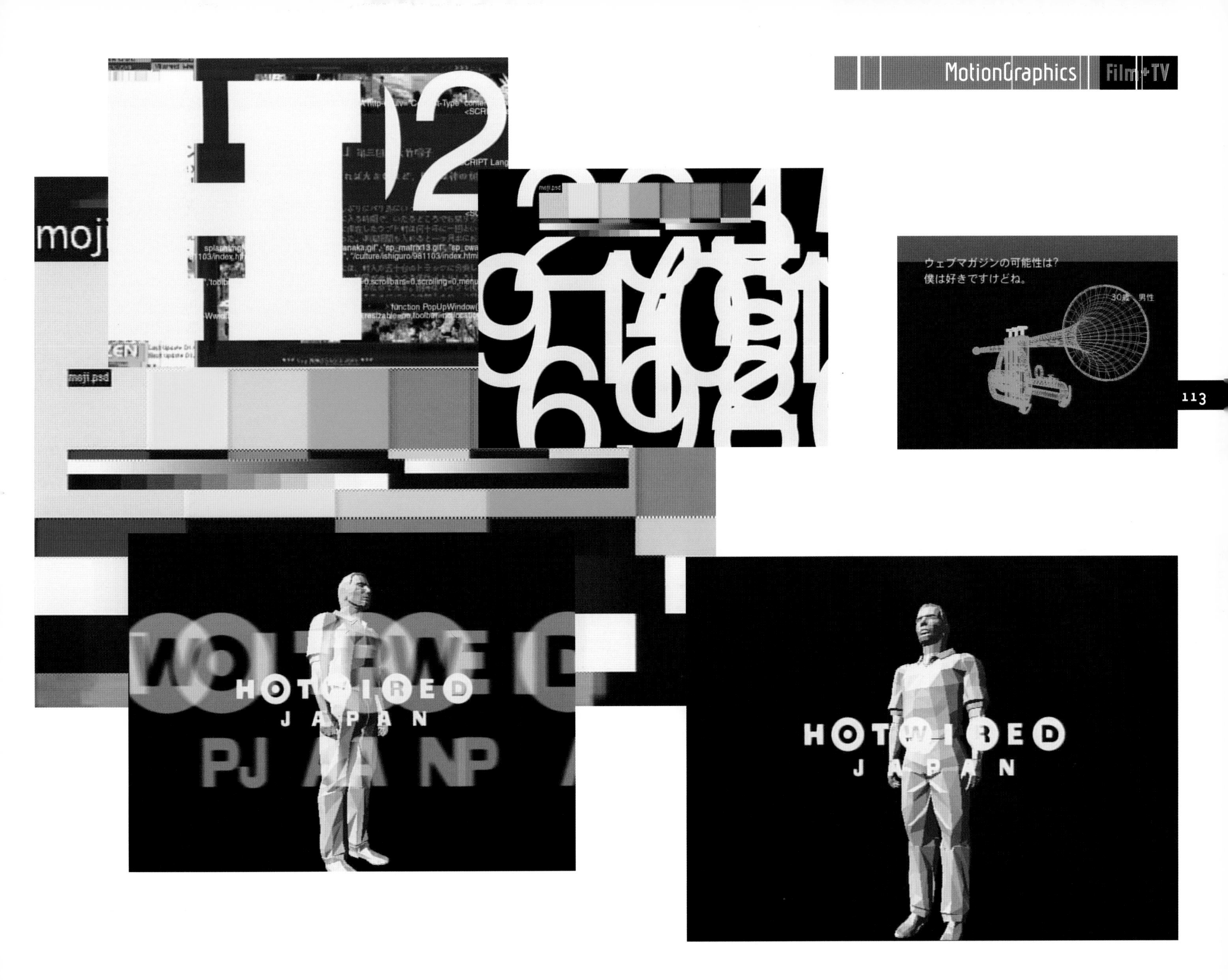
moji
moji.psd
ウェブマガジンの可能性は?
僕は好きですけどね。
30歳、男性
HOTWIRED
JAPAN
HOTWIRED
JAPAN

Viewsic "Live-UK"

Drawing and Manual was commissioned to design the opening title sequence for Viewsic's live television show showcasing the UK-Tokyo musical connection. The concept blends imagery of broadcasting with elements associated with London and Tokyo. A contrasting, yet harmonious pairing of blue and orange dominate the color palette. An interesting effect was achieved by using filmic images of Tokyo, while the London scenes were mainly illustration-based. The designers combined many different cuts with clean and interesting typography to achieve the look of a music scene that correlates Eastern and Western cultures.

Body
Heart beat
Live
The
UK
Live Live Live
Keep on wat
beat
Attra
The Live UK
ASAHI SUPER DRY

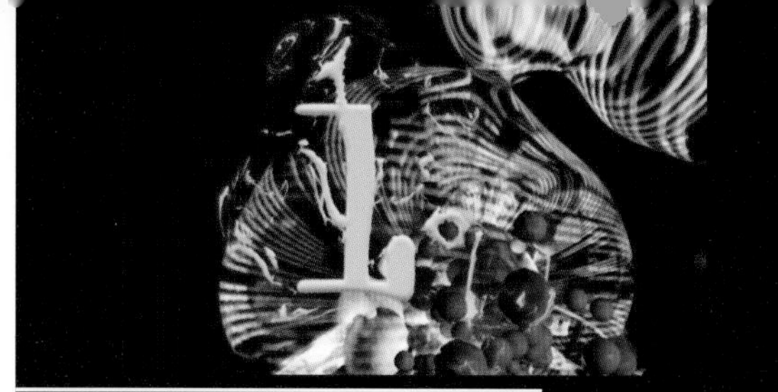

Channel One Liquid Space

The Channel One Network broadcasts to more than 12,000 schools nationwide, presenting news and in-depth features on issues of importance to American teenagers. Channel One asked San Francisco's General Working Group to design a series of thematic show openers. "Liquid Space," one in the series, revolves around a young woman's face at center stage as a means of visualizing telepresence. Morphing shapes and colors move about the girl as she watches video bytes in a virtual world. The onscreen action resolves fluidly into the bright green Channel One logo, reinforcing the network's identity.

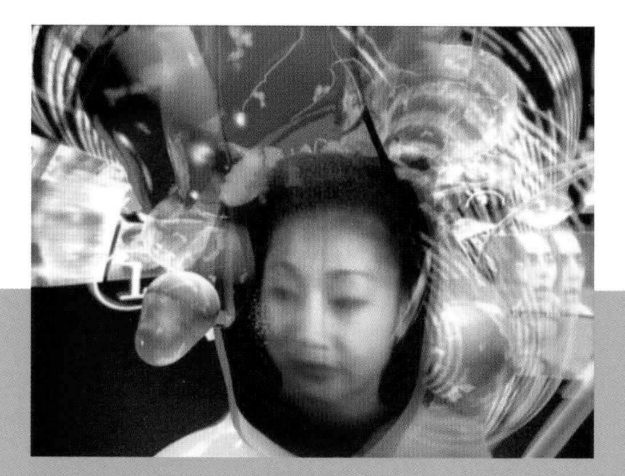

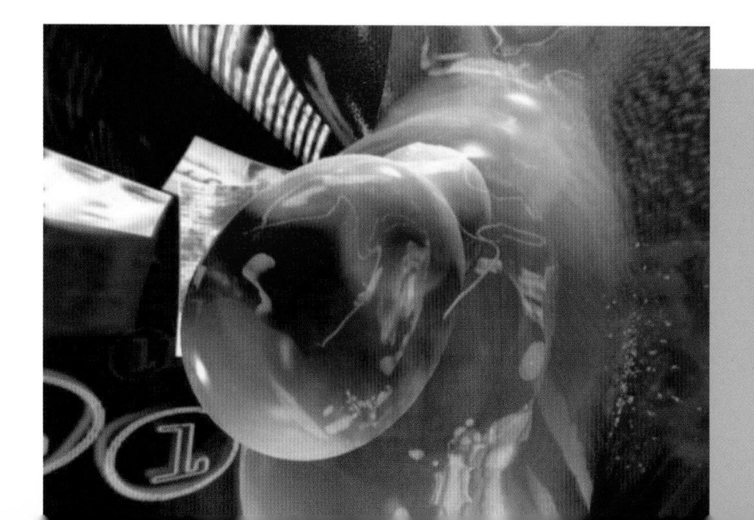

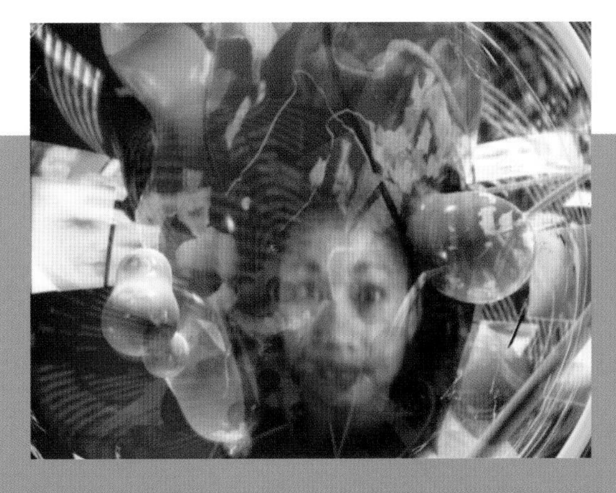

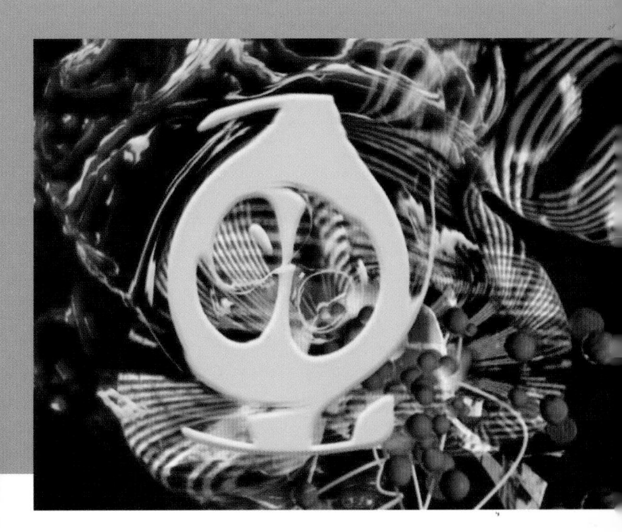

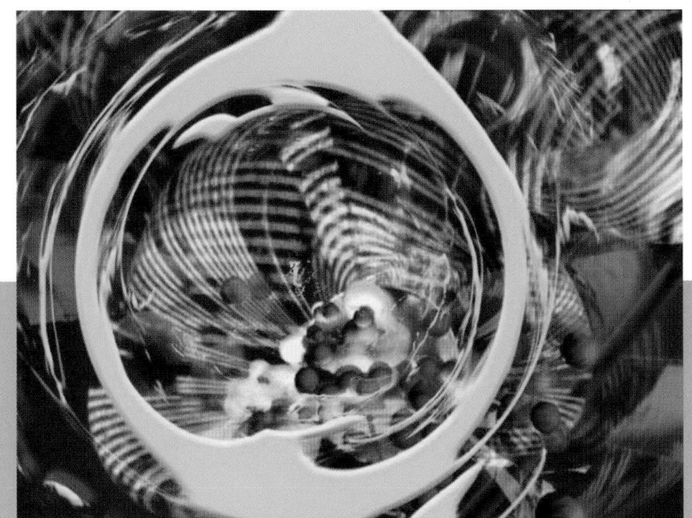

Channel One

Show Open + ID **Channel One** | Designers **Geoff Kaplan, R. Paul Seymour** Studio **General Working Group**

Channel One was investigating the feasibility of producing a current events/news program to appeal to a college-age audience. General Working Group was asked to develop a show package which included the network's identity and show opener. Multiple layers of animated 3D objects combined with filmic imagery tease the viewer as the main typography appears. The C1U logo emerges from the mysterious motion to settle firmly on a cyberspace-style grid horizon, solidly projecting the Channel One identity to its youthful audience.

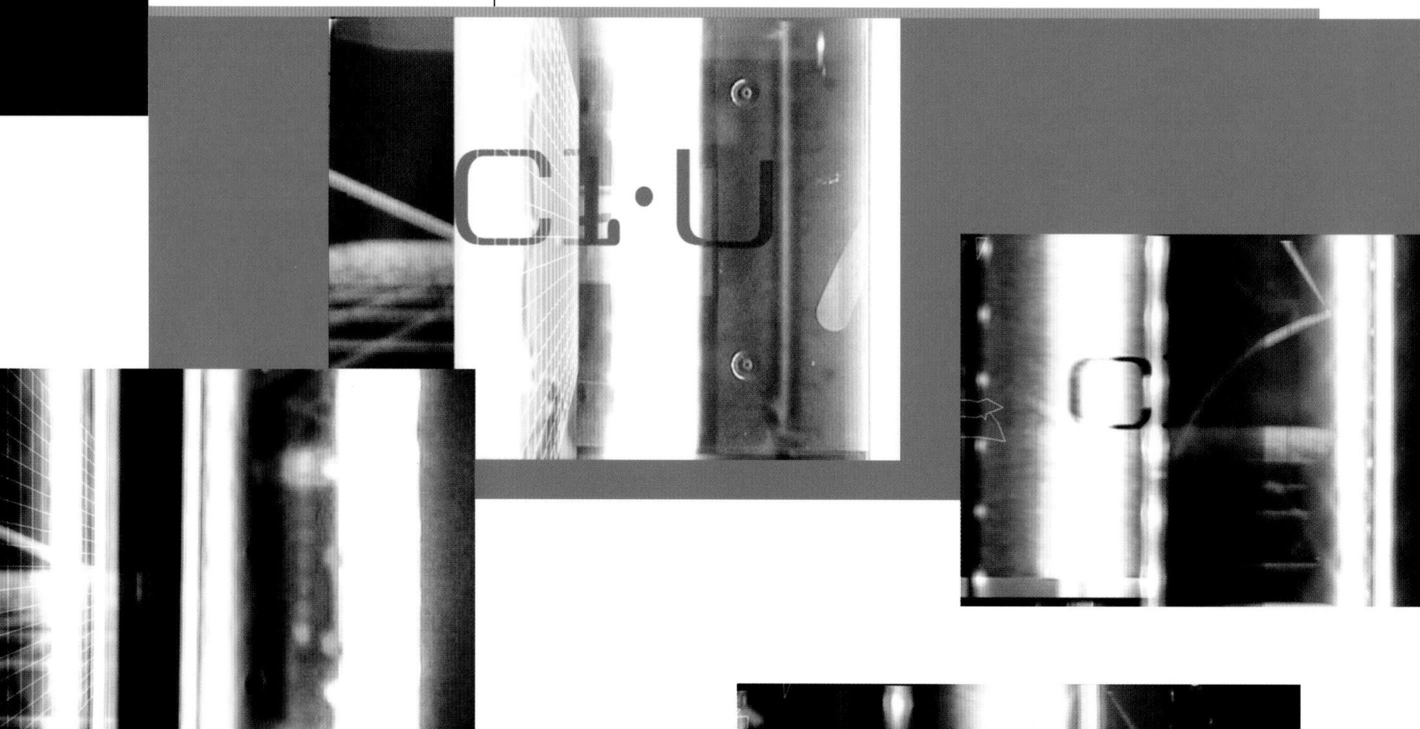

CHANNELONE (UNIVERSITY)

C1·U
CHANNELONE (UNIVERSITY)

Digital Entertainment Network Channel

Digital Entertainment Network Channel Show Open + ID General Working Group designed the motion graphics and show opener for Digital Entertainment Network's reel, geared toward the 14 to 24 year-old global youth market. The piece attracted advertisers and served as identity bumpers proceeding the web shows. The open uses a color scheme of bright green, blue and purple hues as a backdrop for the clean lines of the DEN logo. The combination of high contrast colors, fluid motion and repetition of a simple well-designed logotype, strengthens the emerging network's identity.

StudioGeneral Working Group

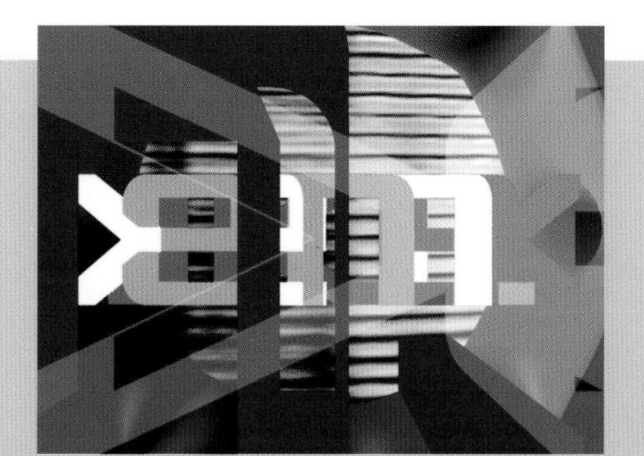

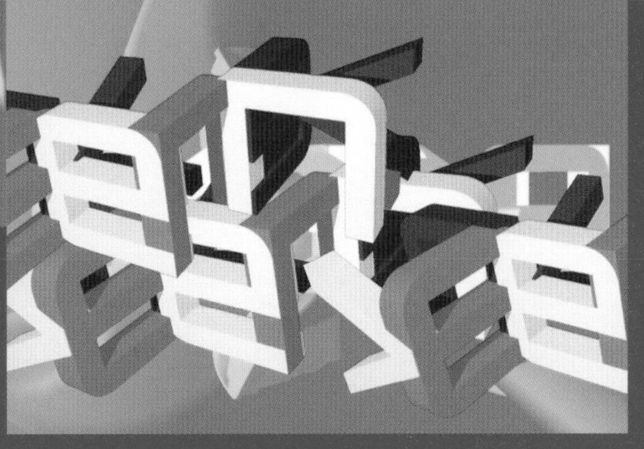

Broadcast Design, Content Library

Vello Virkhaus created the idea for Future Nightlife while directing a visual library for Sign Cast Network. The net's nightlife channel needed a progressive, hyper-real artistic representation of the electronic music and evening entertainment culture. The designers were challenged to create a unique and compelling, yet not too intrusive piece. Initially, they made a 3D library of centric maze-like forms to serve several design tasks. The v2 team also designed a series of morphing shape loops, tentacles, grids and geometric multi-sided, multi-angled planes. After building a substantial library of elements, each artist remixed the imagery and composited in live-action blue screen shots of dance performers.

Nightshade, Ariel Martian, Matt Daly ClientSign Cast Network Studiov2 Labs, OVT Visuals ProductionBrian Dressel, OVT

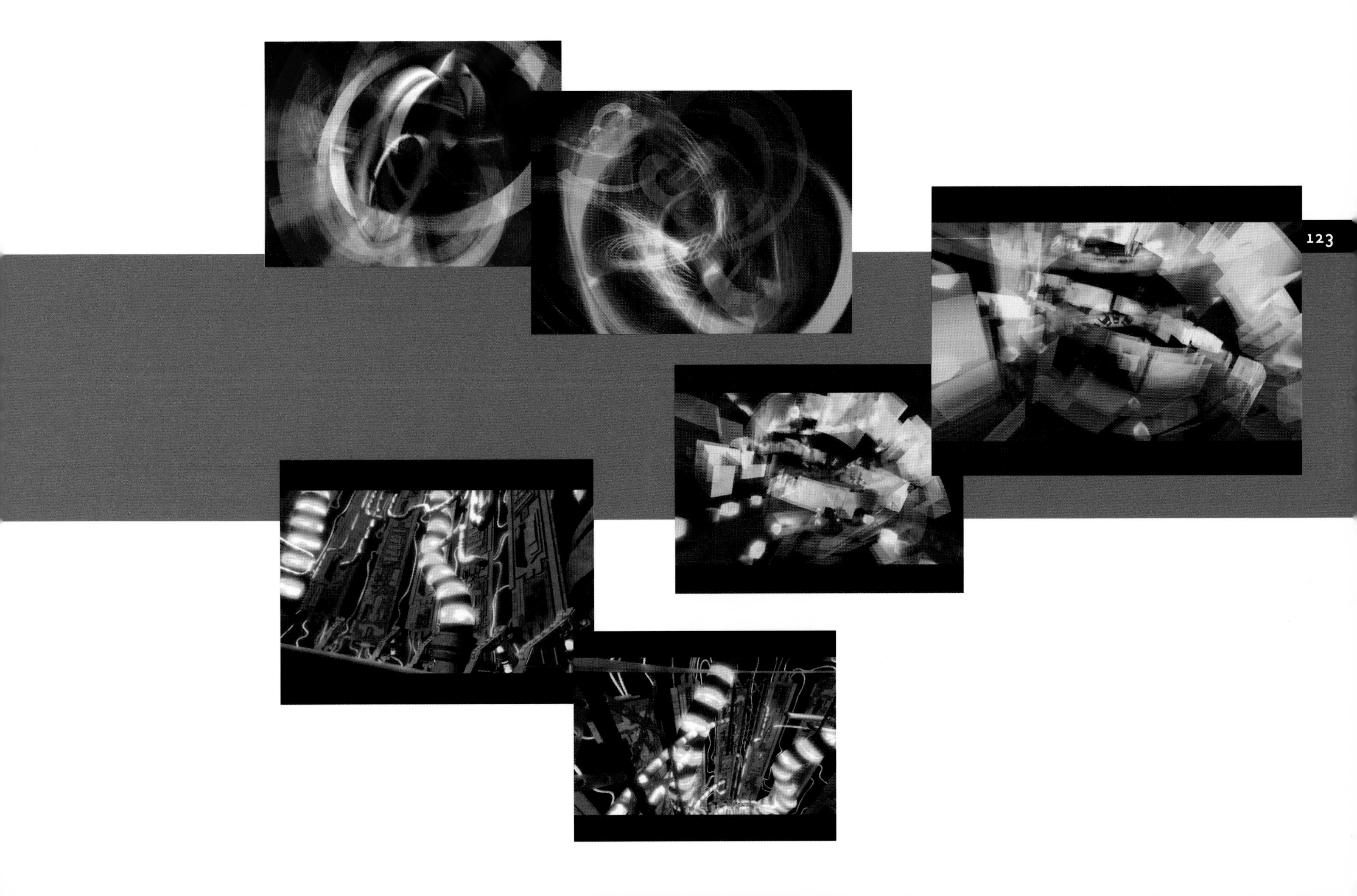

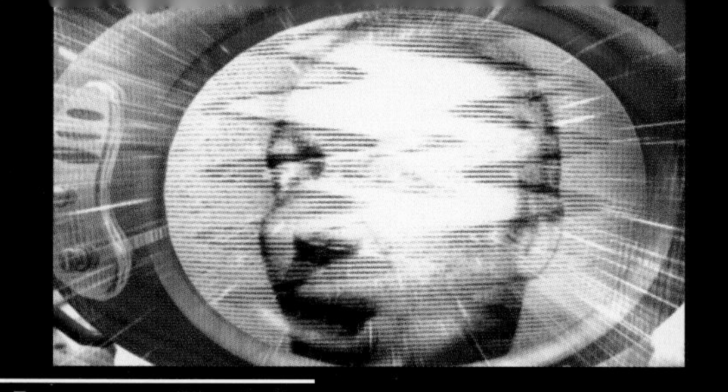

Show Open **Internet Tonight** | Directors **Lisa Rubinelli, Vello Virkhaus, v2** Animation **v2, Lisa Rubinelli,**

Internet Tonight

ZDTV called upon Vello Virkhaus to design a show open for "Internet Tonight," and create a library of broadcast elements for the network's use. During concept stages, Virkhaus discovered an interesting character, "TV Head." The designer took an old 1950s-style round television and integrated it with a holographic computer screen, adding glowing vacuum tubes and tangled wiring. The character became a digital hero, sailing over a sea of internet drones, each with a tiny moving web cam mounted on its head. The show opens with the drones spewing out web URLs, leading the viewer to information as seen through the eyes of TV Head.

Doug Carney, H-Gun Client ZDTV, Internet Tonight Show Studio v2 Labs, H-Gun SF

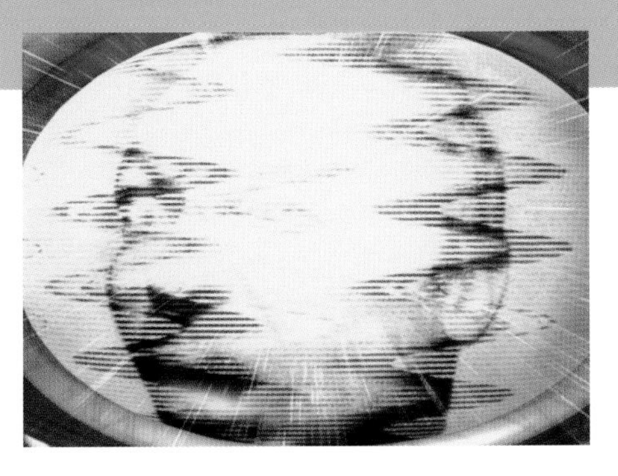

35mm Film Trailer **The Method Festival** | Director **Vello Virkhaus, v2** Animation **v2** Client **The Method Festival LA**

The Method Festival

Vello Virkhaus' challenge was to design a television commercial and film trailer from one animation sequence. The Method Festival required that the piece integrate the existing aesthetic while representing Method acting highlights. Virkhaus pulled over 30 selections from each film, showcasing some of the finest moments from the acting performances. To avoid rendering at film resolution, Virkhaus maintained the ratio of film, but worked at one quarter of the normal resolution. He digitally changed the resolution via fractal enlargement. The finished sequence went from 640x480 up to 2K, maintaining the quality and visual integrity the client demanded.

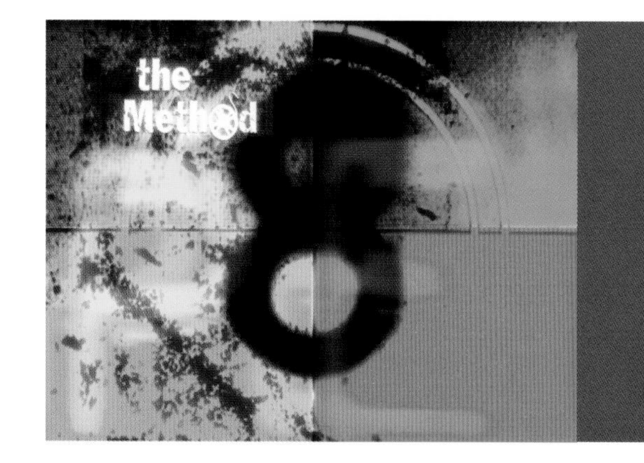

Studiov2 Labs, CF

The Quest for the "Higgs"

The most challenging conceptual project Vello Virkhaus has executed to date is "The Quest for the Higgs." The director and sponsor, Dr. Martinus Veltman, won a Nobel prize in 1999 for his Electroweak theories. Virkhaus worked with Veltman's son Martijn, to create this piece. After numerous hours of research, the designers developed a two minute montage of the history of science, from the first human tool to particle accelerators and the universe's smallest particle, the "Higgs." The remainder of the film, which doubled as a lecture tool, explores the esteemed doctor's work, connecting art and science in a technological fashion.

Martinus Veltman, Arcane Realities Studio**v2 Labs**

Reel Open **Belief Reel** | Designer **Belief** | Client **Internal**

Belief Reel

Communication creates culture. Culture evolves the sensory experience. The sensory experience defines communication. This infinite loop creates your Belief. Immerse in Sensory Culture is the mantra for this movement. Belief's approach to nurturing creativity is unique. By bringing together fine artists, illustrators, musicians and filmmakers, Belief provides an unparalleled creative resource. The Belief culture encourages varying talents, strengths and temperaments to influence a diverse approach to each project. The studio aims to architect the new sensory culture, making communication a more fluid and compelling experience by marrying multiple creative disciplines.

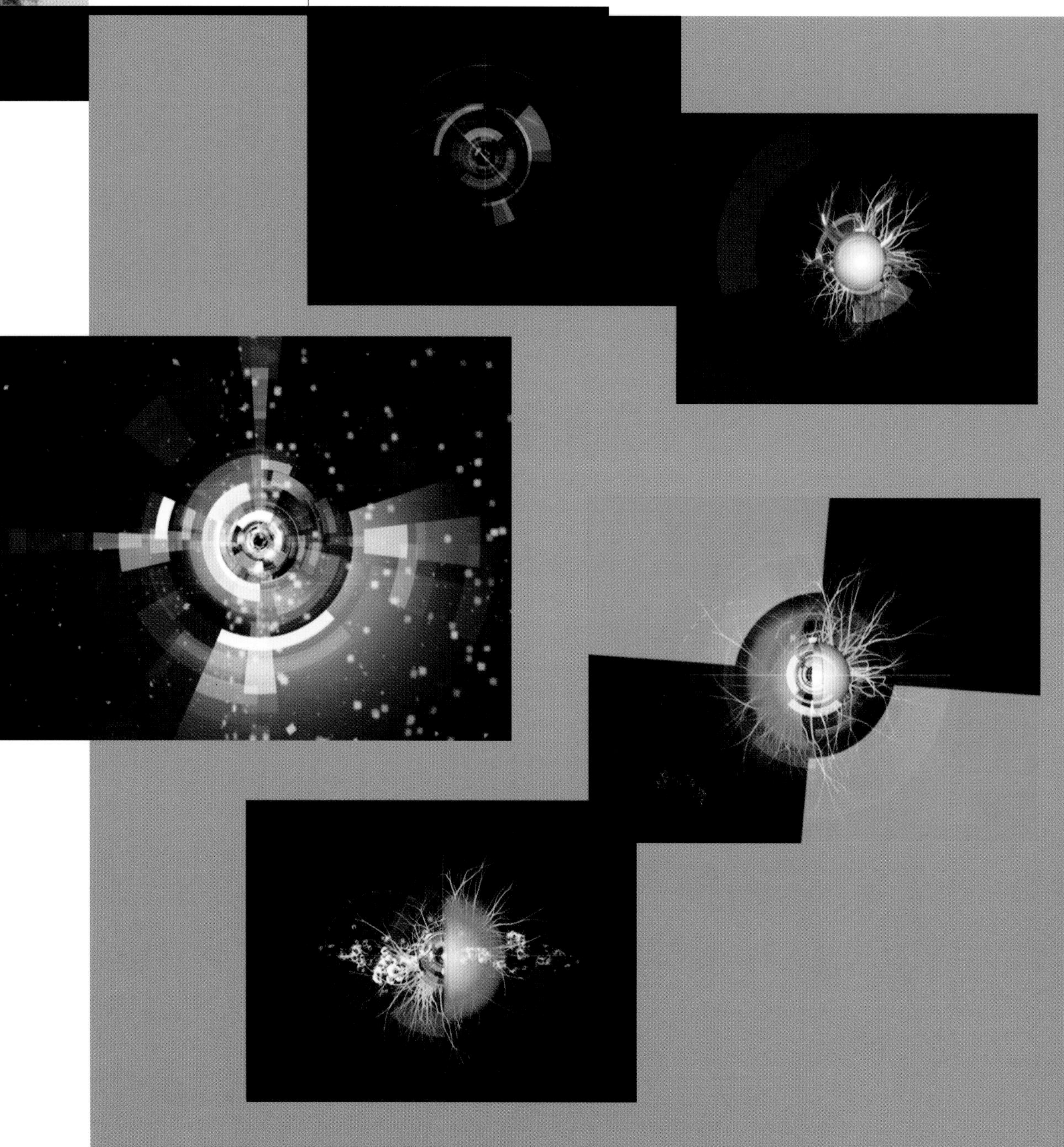

belief
belief
belief
belief
Immerse In Sensory Culture
310 - 998 - 0099
www.belief.com
belief

Alive Network IDs

Alive Belief created a Network Launch for Alive; a global network aimed towards moderate income jet-set travelers. Working with the premise of "attainable travel," liquids were shot by the studio and composited to create fantastic landscapes. People were shot on greenscreen and then silhouetted to keep the package culturally and ideologically independent and work in any region of the world in any language. The creative toolset was Media100, Adobe Photoshop, Adobe Illustrator and Adobe After Effects.

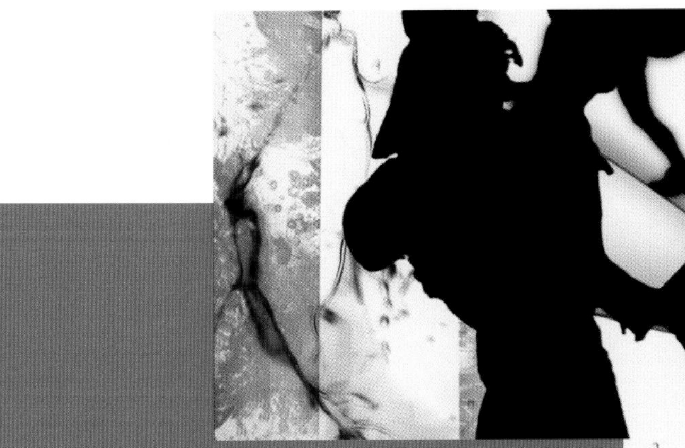

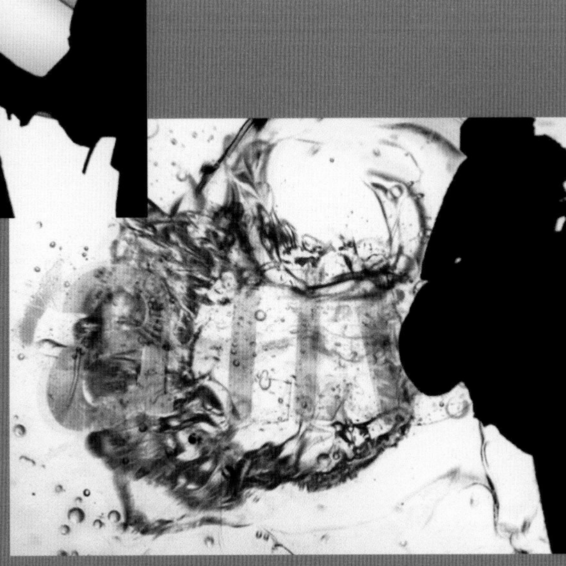

alive

alive
NETWORKS

alive
NETWORKS

Modern Man

Modern Man, created by Belief, begins with an amazing audio recording of an infamous Venice boardwalk character known as Preacher Tom. Vintage stock footage combined with graphics tells Preacher Tom's story about infinity. Vast landscapes of digital clutter were made to represent the speakers brain using Adobe After Effects 3D features. Belief designers, working independently, created various animations. The final graphics were edited in Media 100 to sound design created by music house Barton: Holt.

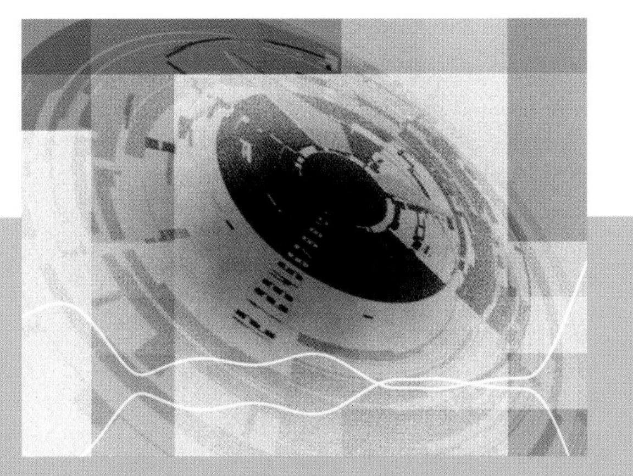

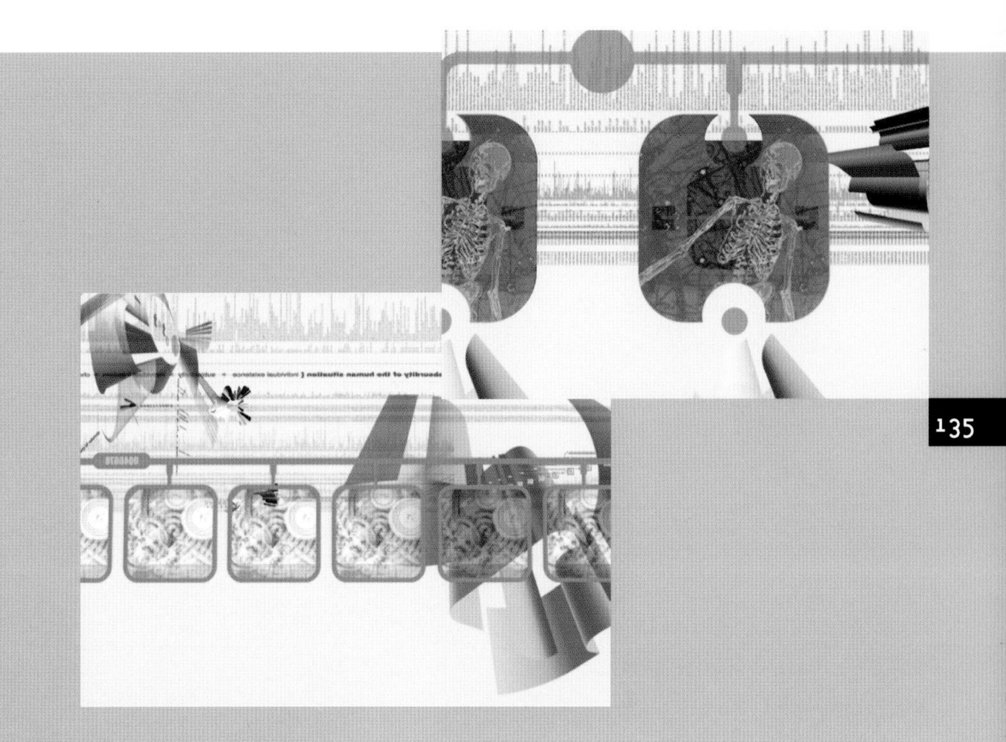

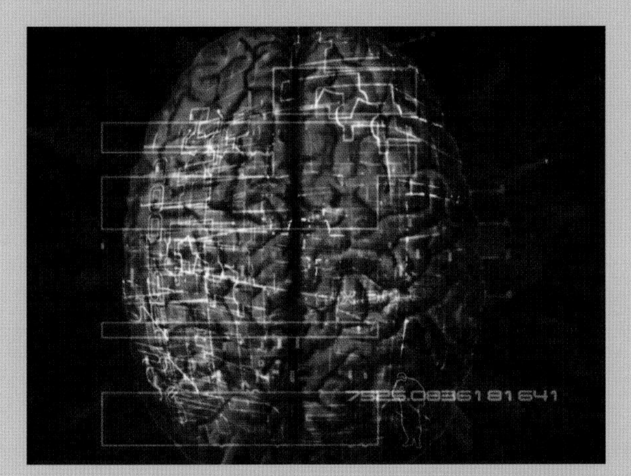
7525.08361 81 641

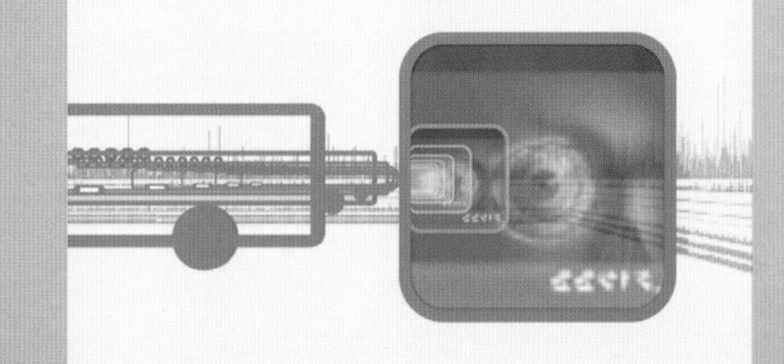

Audiofile

Tech TV asked Belief to create the show package for Audiofile, Tech TVs premiere new television series about online and digital music. The goal was to create a friendly yet hi-tech environment to showcase people creating and enjoying music. Belief built large Lucite collages of bizarre electronic devices and parts. 35 mm live action footage of dancers and musicians were edited together using Media 100 and then tracked and composited into the piece using Adobe After Effects. Belief created additional graphic elements to give the piece an interactive feeling. The final showpackage toolkit included transitions, backgrounds and bumpers.

AUDIOFILE

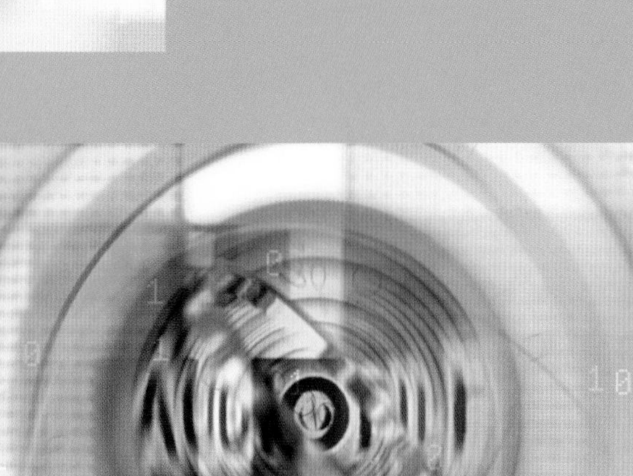

Belief EXP Reel

Belief EXP was founded in 1999 by the artists at Belief to stretch the talents of its in-house designers in the arena of experimental media, fine art and noncommercial creative content. Belief EXP is more than an arm of the studio, it is a philosophy and an open-ended forum for the flow of vanguard ideas, communication exploration and the development of progressive processes and techniques. Belief EXP is the artistic medium for the creation of experimental content. This open for the Belief EXP reel was created by shooting water bubbles with a microscope. Belief used Media 100 and Adobe After Effects to edit and composite the elements and logo.

belief
EXP
belief
EXP
EXP

Smooth Warming

Belief dived into the art world with this experimental design piece. Working with the most difficult imagery imaginable (unicorns and mermaids) Belief attempted to make these fairytale icons into something modern. The goal of the piece was to create a feeling of an animated painting. Experimenting with process algorithms, plug-ins and scripting, Belief used Adobe After Effects to create 30 minutes of graphics for this LA MOCA video installation.

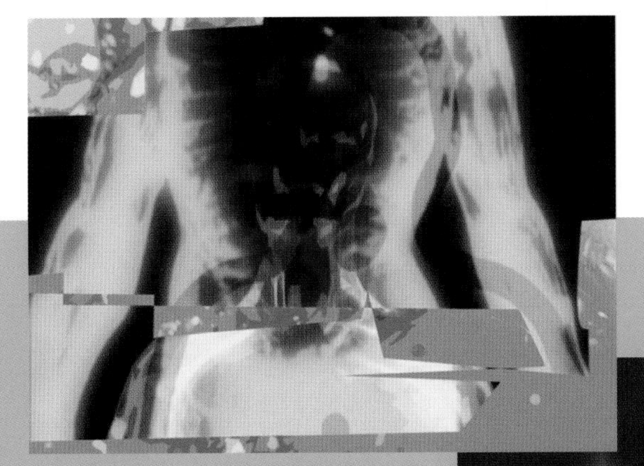

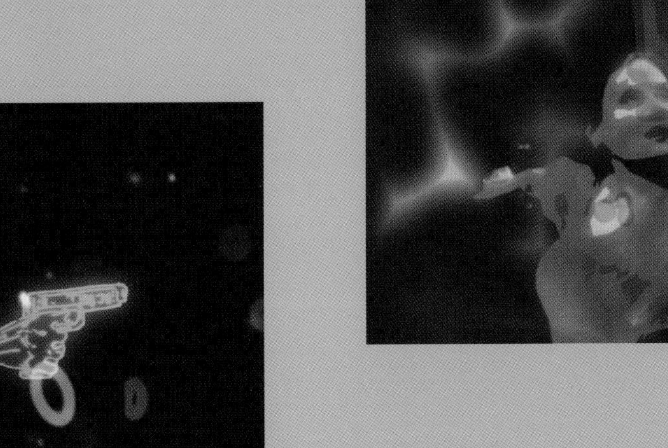

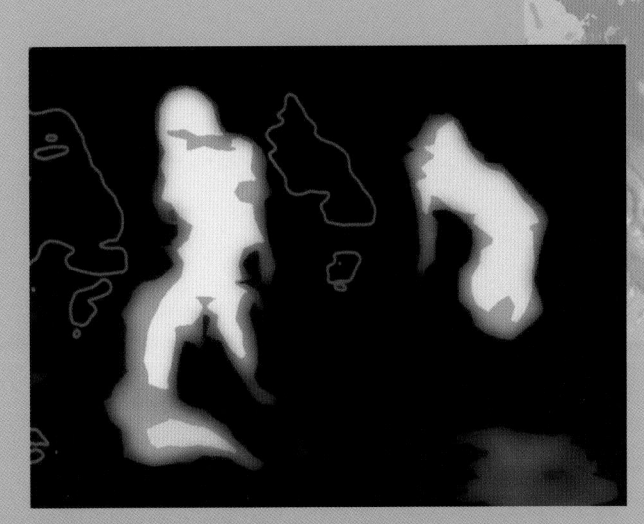

So You Wanna Be A TRL Superstar: Debuts and Demo Tapes

For MTV's "So You Wanna Be A TRL Superstar: Debuts and Demo Tapes," Belief created a huge layout of microphones, cables and exotic monitors. The "sculpture" was photographed in sections with a hi-resolution digital still camera. These hi-res sections were joined in Adobe Photoshop to create a larger image that was brought into Adobe After Effects and animated. The large file size allowed Belief to zoom into the images without loosing resolution. Digital "schmutz" and typography were added to the piece in the final composites. The result influenced not only the opening of the show but the set design as well.

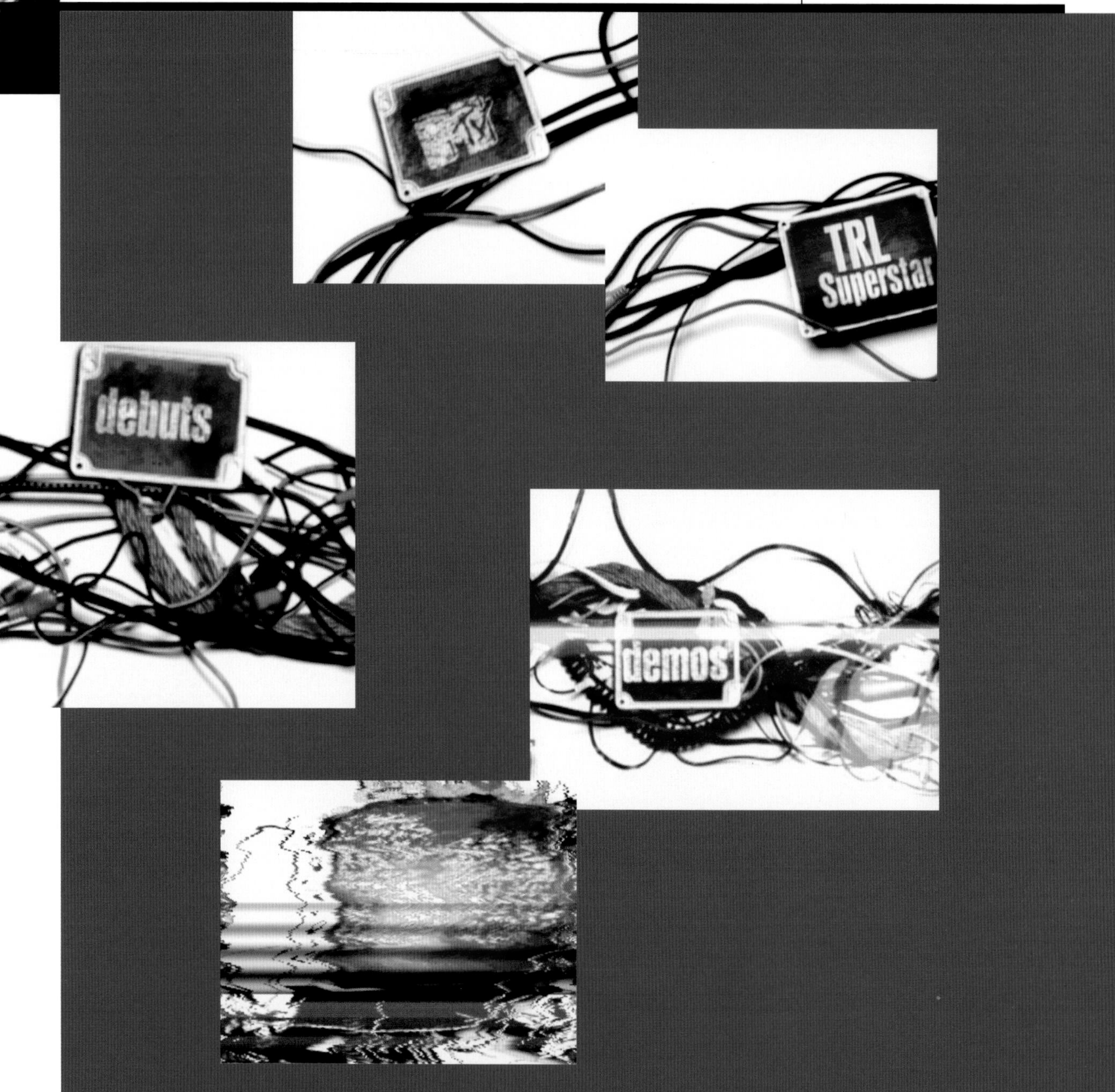

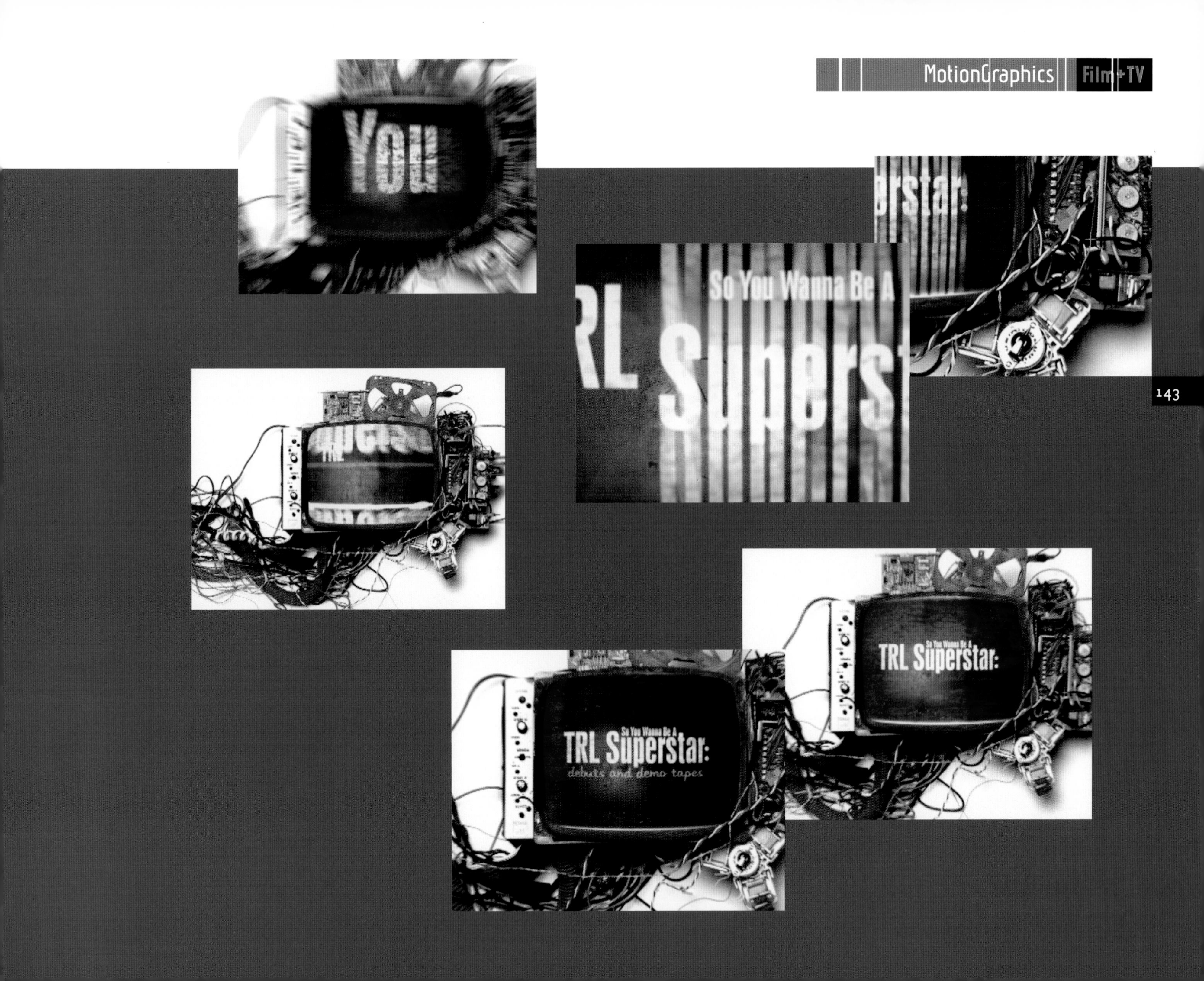
You
So You Wanna Be A
TRL Superstar:
debuts and demo tapes

Honda Power of Dreams

Le Cabinet was called upon to develop a 30-second introductory film for Honda's European Pan Convention in September 2001. The piece was based on a count-down and structured as a virtual ride through diverse 3D land-scapes, including a race track, mountain road and cityscape. The subjective movements of the camera are purposefully fast and nervous, evoking for the audience the singular thrill of speeding along on a fast and powerful motorbike.

Music **Jean-Louis Morgère** Studio **Le Cabinet**

Network TV Branding **Flix Event** | Client **Showtime, New York** Senior Creative Director **Ann Weiser** Creative Director **New York** Creative Directors **Jakob Trollbäck, Antoine Tinguely** Art Directors **Laurent Fauchere, Nathalie de La Gorce**

Flix Event

Showtime asked Trollbäck to create a comprehensive branding package for the Flix network. The client provided its logo typeface and four-color palette, along with the tagline and an outline of the programming for each of the Flix franchises. Flix Event is a theme for a night of programming, showcasing films that feature a certain actor, director, genre, etc. The color orange was selected for its sense of urgency. Since the common thread for all the franchises is film, Trollbäck broke the design down to the medium's most basic element: light. Lighting effects were inspired by the illumination in a Xerox scanner, the lights of the New York subway system and glowing neon that travels, leaving tracers of light in its wake.

Anthony Castellano Art Director**Christina Black** Writer/Producer**Erik Friedman** Production Managers**Lorraine O'Connor, Diana L. Roach** Design Firm**Trollbäck & Co.,** Producer**Meghan O'Brien** Sound Design**Sacred Noise, New York** Composers**Pete Rundquist, Chuck Lovejoy, Dave Gennaro**

Network TV Branding **Flix Pick** Client **Showtime, New York** Senior Creative Director **Ann Weiser** Creative Director **New York** Creative Directors **Jakob Trollbäck, Antoine Tinguely** Art Directors **Laurent Fauchere, Nathalie de La Gorce**

Flix Pick

To differentiate Showtime's Flix franchises, Trollbäck and Company used a different primary color for each feature lead. They also developed a different light source shape for each franchise. The Flix Pick represents the network's "pick of the month." Red was selected because it was part of the client's identity, and Trollbäck and Company wanted each franchise to have its own palette while still referencing the brand's identity. The lighting for Flix Pick was inspired by the kind of spotlights one would see at a major Hollywood motion picture premiere.

Anthony Castellano Art Director**Christina Black** Writer/Producer**Erik Friedman** Production Managers**Lorraine O'Connor, Diana L. Roach** Design Firm**Trollbäck & Co.,** Producer**Meghan O'Brien** Sound Design**Sacred Noise, New York** Composers**Pete Rundquist, Chuck Lovejoy, Dave Gennaro**

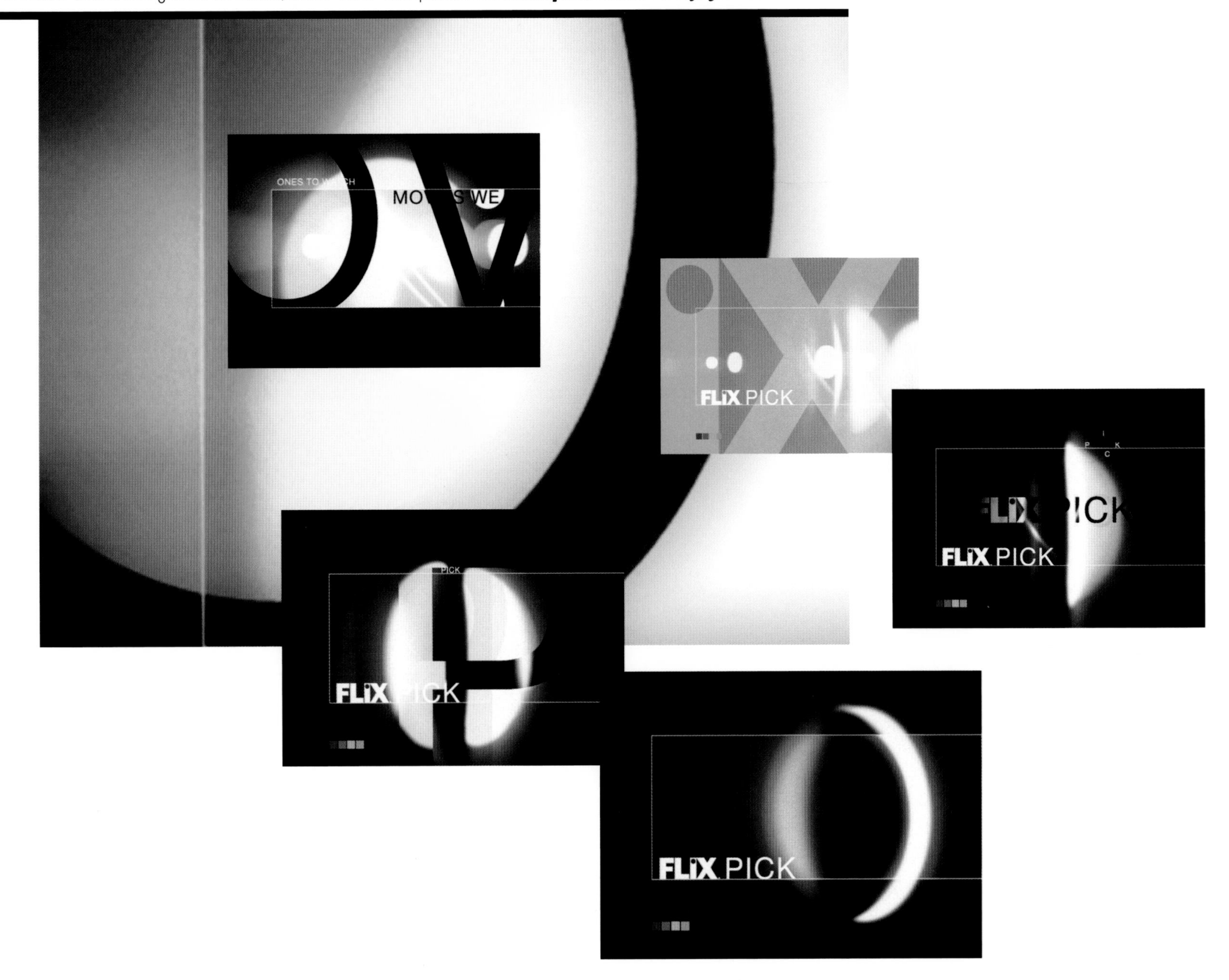

Network TV Branding **Flix 5os** Client **Showtime, New York** Senior Creative Director **Ann Weiser** Creative Director **New York** Creative Directors **Jakob Trollbäck, Antoine Tinguely** Art Directors **Laurent Fauchere,**

Flix 5os

The greatest challenge was Trollbäck's vision calling for lighting effects that are normally achieved by shooting on film. Because of time and budget restrictions, the creative team generated all of the design effects by experimenting in Adobe After Effects. For the Flix 5os, project, Trollbäck aimed for a nostalgic mood, employing gray because it felt retro, and was reminiscent of the period's black-and-white movies. Trollbäck also used lens flares in the lighting, achieving the hexagonal effect that occurs when light hits the camera directly.

Anthony Castellano Art Director**Christina Black** Writer/Producer**Erik Friedman** Production Managers**Lorraine O'Connor, Diana L. Roach** Design Firm**Trollbäck & Co., Nathalie de La Gorce** Producer**Meghan O'Brien** Sound Design**Sacred Noise, New York** Composers**Pete Rundquist, Chuck Lovejoy, Dave Gennaro**

Network TV Branding **Flix Cool Classics** | Client **Showtime, New York** Senior Creative Director **Ann Weiser** Creative Director **New York** Creative Directors **Jakob Trollbäck, Antoine Tinguely** Art Directors **Laurent Fauchere, Nathalie de La Gorce**

Flix Cool Classics

Trollbäck's work for Flix was an attempt to replicate light and its behavior in the real world—how the eye sees light, and the inherent latency involved in moving light and how it flares. Great results often came from creative accidents, which were fleshed out and polished for the final designs. The Flix Cool Classics branding combined elements from the Flix Events, Flix Pick and Flix 50s projects. Trollbäck used circles taken from the "I" of the Flix logo and square windows to bring together a piece that was related, yet maintained its own identity. The Cool Classics' orange, red, gray, blue and purple hues reference the Flix palette, reinforcing the network's identity.

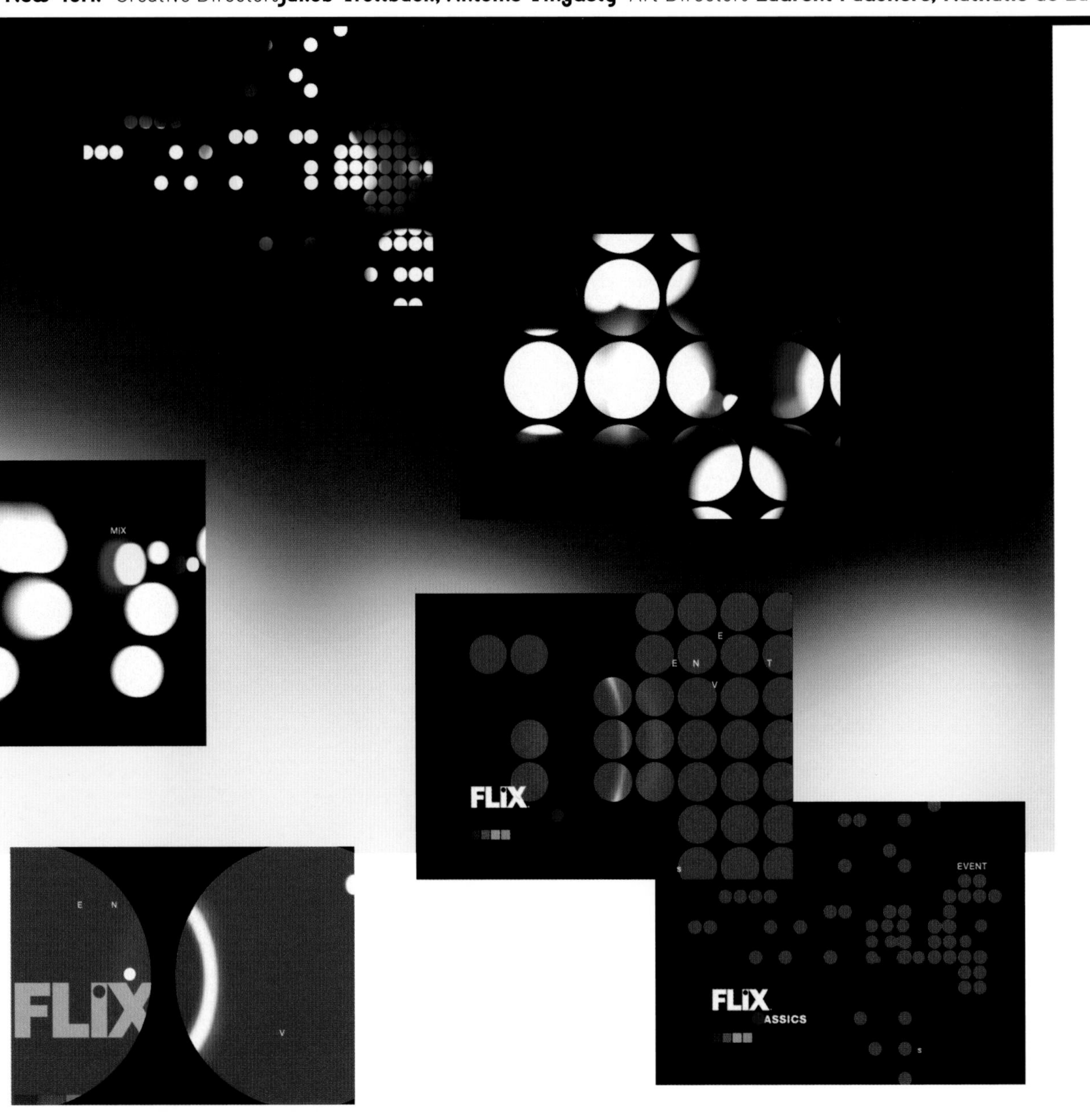

Anthony Castellano Art Director**Christina Black** Writer/Producer**Erik Friedman** Production Managers**Lorraine O'Connor, Diana L. Roach** Design Firm**Trollbäck & Co.,** Producer**Meghan O'Brien** Sound Design**Sacred Noise, New York** Composers**Pete Rundquist, Chuck Lovejoy, Dave Gennaro**

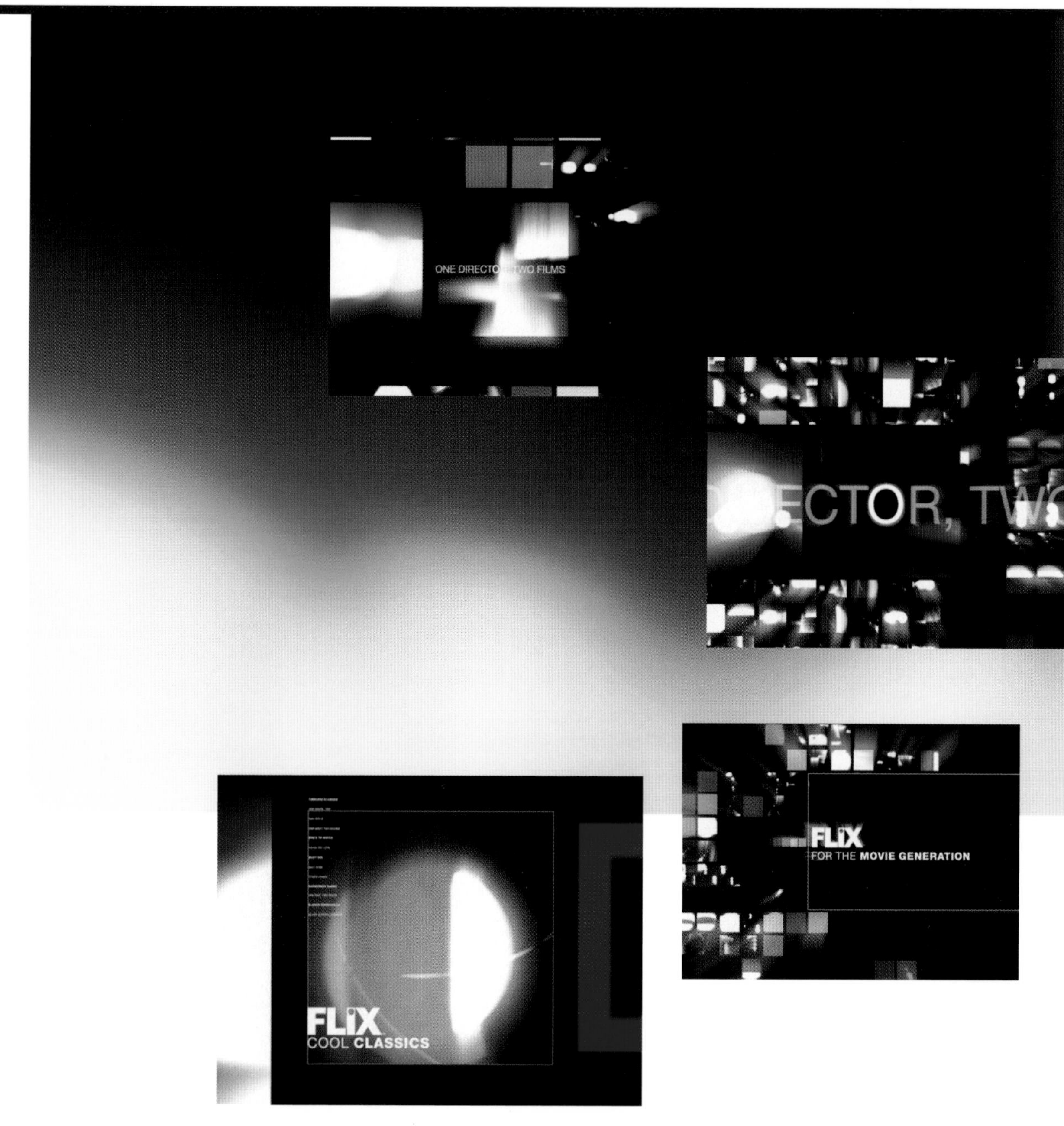

JUNE FOR SAVING.

TV Commercial**Volvo Safety** Client**Volvo Cars (a division of Ford Motor Company)** Executive**Peter Wood Trollbäck & Company** Creative Director**Jakob Trollbäck** Creators/Directors**Antoine Tinguely, Laurent Fauchere** CGI Artist

Volvo Safety

"Safety Pin," designed and produced in-house, introduced a crisp, new signature branding style for the Swedish car manufacturer. The concept was simple: use spare, intelligent design to communicate the Volvo engineering philosophy while cutting through commercial television clutter. The spots, which integrate photography featured in Messner's Volvo print campaign, are based on simple, energetic graphics powered by heavy rhythmic music tracks. Trollbäck and Company creative directors Antoine Tinguely and Laurent Fauchere, both from Switzerland, saw the spots as an exercise in getting maximum impact from the purest possible graphic language. "It's wonderful when you work with an agency team that understands that if you have a great product, the message should be straight and undiluted," says Fauchere.

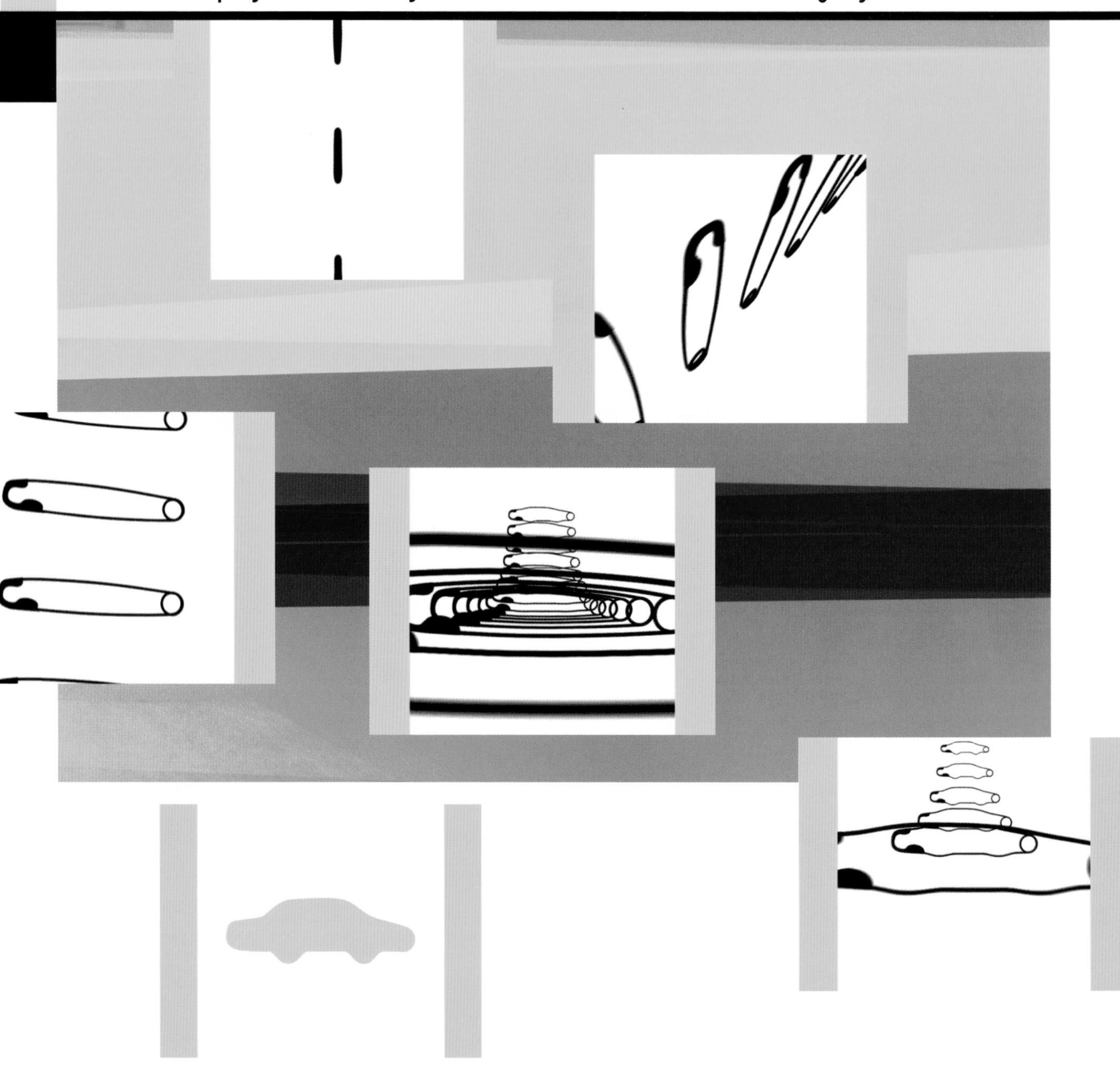

Agency **Messner Vetere Berger McNamee Schmetterer/EURO RSCG and FUEL North America** Agency Producer**Sara Cole** Graphic Design Facility **Chris Haak** Editor**Nicole Amato** Senior Producer**DeDe Sullivan** Associate Producer**Tricia Chatterton** Music Design Company**Sacred Noise, New York** Studio**Trollbäck & Company**

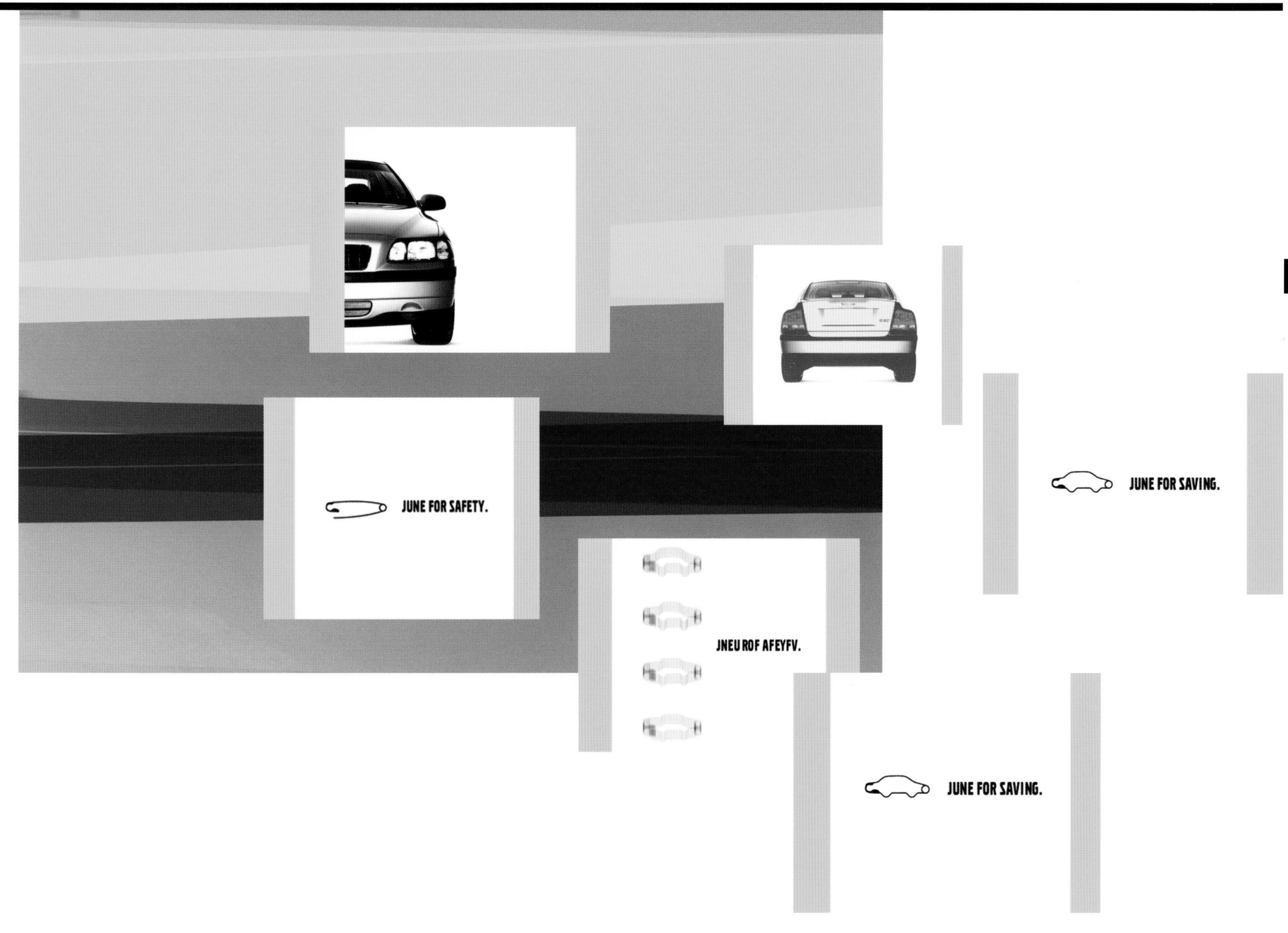

Film Title **Hysterical Blindness** | Designer Early Phase Storyboarding **Jasmin Jodry**

Director **Mira Nair** Designer **Chris Haak** Studio **Trollbäck & Company** Production Company **HBO Films**

Hysterical Blindness

Hysterical Blindness was the second film title sequence Trollbäck and Company created for director Mira Nair. Nair's idea was to open the movie with an eye exam, as Uma Thurman's character is examined for a spontaneous case of blindness. An eye chart was used for the main title card, with light becoming another key element. Used as a transitional device, in and out-of-focus shots from Thurman's point of view were created to evoke the feeling of waking from a deep sleep before your eyes have adjusted to bright light. Trollbäck creatives attended the shoot and co-directed the sequence with Nair. Additional footage was shot as needed to achieve the light source effects.

Freelance Designer Storyboarding **Greg Hahn** Creative Directors **Laurent Fauchere, Jakob Trollbäck, Antoine Tinguely**

Belief 130–143
Mike Goedecke, Principal / Executive Creative Director
1832 Franklin Street
Santa Monica, CA 90404
United States
Phone: 310.998.0099
Fax: 310.998.0066

Email: goedecke@belief.com
Website: http://www.belief.com

BL:ND 72–79
830 Colorado Avenue
Santa Monica, CA 90401
United States
Phone: 310.314.1618
Fax: 310.314.1718
Email: design@blind.com
Website: http://www.blind.com

D-FUSE 104–109
3rd Floor / 36 Greville Street
London EC1N 8TB
United Kingdom
Phone: +44(0)20.7419.9445
Email: claire@dfuse.com
Website: http://www.dfuse.com

DRAWING AND MANUAL 110–115
8-3-2 Okusawa, Setagaya
Tokyo 158-0083 Japan
Phone: +81.3.5752.0097
Fax: +81.3.3705.0377
Email: info@d-a-m.co.jp
Website: http://www.d-a-m.co.jp

Engine 40–45
Finnegan Spencer
47 Herbert St Artarmon
Sydney NSW 2064
Australia
Phone: 612.9438.2000
Fax: 612.9439.5957
Email: finn@engine.net.au
Website: http://www.engine.net.au

evolution | bureau 26–27
Jason Zada
90 Tehama Street
San Francisco, CA 94105
United States
Phone: 415.243.0859
Fax: 415.243.9255
Email: Jason@evolutionbureau.com
Websites: http://www.jasonzada.com
http://www.evolutionbureau.com

FLUX group 30–33
Joseph Kiely
215 West Superior Street / 7th floor
Chicago, IL 60610
United States
Phone: 312.640.1111
Fax: 312.640.1018
Email: kiely@fluxgroup.com
Website: http://www.fluxgroup.com

General Working Group 116–121
Geoff Kaplan
1153 Florida Street
San Francisco, CA 94110
United States
Phone: 415.648.4617
Email: geoff@generalworkinggroup.com
Website: http://www.generalworkinggroup.com

Graphic Havoc avisualagency 100–101
184 Kent Avenue / #311
Brooklyn, NY 11211
United States
Phone: 718.388.5519
Fax: 718.387.7383
Email: info@graphichavoc.com
Website: http://www.graphichavoc.com

Gratex International 22–25
Drienova 3
Bratislava 821 02 Slovak Republic
Phone: +421.2.434.294.90
Fax: +421.2.434 .101.18
Email: sales@gratex.com
Website: http://www.gratex.com

GRUPO DOMA 102–103
Buenos Aires, Argentina
Email: info@doma.tv
Website: http://www.doma.tv

Head Gear Animation 98–99
Julian Grey
35 McCaul Street / Suite 301
Toronto ON M5T 1V7 Canada
Phone: 416.408.2020
Fax: 416.408.2011
Email: info@headgearanimation.com
Website: http://www.headgearanimation.com

Hatmaker 48–55
Chris Goveia, Partner/Creative Director
63 Pleasant Street
Watertown, MA 02472
United States
Phone: 617.924.2700
Fax: 617.924.0019
Email: cgoveia@hatmaker.com
Website: http://www.hatmaker.com

Le Cabinet 144–145
Marc Nguyen Tan + Benoit Emery
6, rue Amélie
Paris 75007 France
Phone: 00.1.44 .18.32.34
Fax: 00.1.44.18.32.34
Email: infos@dotmov.com
Website: http://www.dotmov.com

The Light Surgeons 34–35
Chris Allen
Unit 409, 134-146 Curtain Road
London EC2A 3AR
United Kingdom
Phone: +44(0)207.613.5756
Fax: +44(0)207.613.5756
Email: info@thelightsurgeons.co.uk
Website: http://www.thelightsurgeons.co.uk

Motion Theory 46–47
Mathew Cullen
1337 Abbot Kinney Blvd.
Venice, CA 90291
United States
Phone: 310.396.9433
Fax: 310.396.7883
Email: javier@motiontheory.com
Website: http://www.motiontheory.com

OVT Visuals, Inc. 10–17, 36–39, 122–123
Brian Dressel
4130 North Claremont Street
Chicago, IL 60618
United States
Phone: 773.583.9344
Email: brian@ovtvisuals.com
Website: http://www.ovtvisuals.com

Planet Propaganda 28–29
Kevin Wade
605 Williamson Street
Madison, WI 53703
United States
Phone: 608.256.0000
Fax: 608.256.1975
Email: hank@planetpropaganda.com
Website: http://www.planetpropaganda.com

StyleWar 66–69
Linnégatan 51
Stockholm 114 58 Sweden
Phone: +46.8.662.8890
Fax: +46.8.662.8895
Email: we@stylewar.com
Website: http://www.stylewar.com

Trollbäck & Company 146–157
302 Fifth Avenue
New York, NY 10001
United States
Phone: 212.529.1010
Fax: 212.529.9540
Website:http://www.trollback.com

twenty2product 18–21
San Francisco, CA
United States
Email: info@twenty2.com
Website: http://www.twenty2.com/

Ultraviolet LLC 56–65
833 Broadway / 2nd Floor
New York, NY 10003
United States
Phone: 212.375.9054
Fax: 212.460.9169
Website: http://www.uv.tv

unburro 70–71
Miguel Rodriguez
4040 NE 2nd Avenue / #309
Miami, FL 33137
United States
Phone: 305.573.4400
Fax: 305.573.4543
Email: info@unburro.com
Website: http://www.unburro.com

Vello 122–129
Vello Virkhaus (AKA VJ v2, v2 animation)
v2
5216 N. Cleon Avenue
North Hollywood, CA 91601
United States
Phone: 818.694.6068
Email: info@vellov.com
Website: http://www.vellov.com

Verb. 84–97
285 West Broadway / Suite 400
New York, NY 10013
United States
Phone: 212.226.9990
Fax: 212.226.0122
Email: info@verbmedia.com
Website: http://www.verbmedia.com

yU+co. 80–83
Garson Yu
941 North Mansfield Avenue
Hollywood, CA 90038
United States
Phone: 323.606.5050
Fax: 323.606.5040
Email: yu_co@yuco.com
Website: http://www.yuco.com